5012
SP-1

Southern Pacific Steam

1940-1957

Joseph A. Strapac

Published 2011 by the
Southern Pacific Historical and Technical Society
Upland, California

Southern Pacific Steam 1940-1957

An annotated photo album of the steam locomotives of SP's Pacific Lines from the beginning of World War II until final dieselization in 1957.

Printed in the United States of America

Publisher's Cataloging-in-Publication
(Provided by Quality Books, Inc.)

Strapac, Joseph A.
Southern Pacific Steam : 1940-1957 / Joseph A. Strapac.
p. cm.
Includes bibliographical references.
LCCN 2010942680
ISBN-13: 978-0-9846247-1-3
ISBN-10: 0-9846247-1-6

1. Southern Pacific Railroad Company. 2. Steam locomotives--United States.
3. Railroads--United States.
I. Southern Pacific Historical & Technical Society. II. Title.

TJ603.3.S6S77 2011 625.26'1'0973
QBI10-600256

For Martha.

Published By:

Southern Pacific Historical & Technical Society
1523 Howard Access Road, Suite A
Upland, California 91786-2582

Book design, layout and typography by Joseph A. Strapac.
Set in Adobe Caslon Pro.

Table of Contents

Acknowledgements

Most of the images used in this book were obtained secondhand or third hand; unfortunately, many did not arrive with any attribution to the original photographer. *When the actual photographer was unknown, no photographer credit appears in the caption.* I suspect that most of the "unknown" images were printed by Guy Dunscomb. The known photographers/collectors who supplied the great majority of the attributed photos include Allan Youell, Arnold Menke, Gerald M. Best, Guy L. Dunscomb, Don H. Roberts, John E. Shaw, Jr., W. C. Whittaker, Charles Givens, Douglas S. Richter, Stan Kistler, Phillips C. Kauke, Alan Miller, Norman W. Holmes and Arthur L. Lloyd. Other known photographers or collectors are credited with their photos.

Advice and support has been generously supplied by Douglas Richter, Gordon Ramsey, Charles Givens and Arnold Menke. My wife, Martha, graciously proofread the entire book, but errors of fact or grammar belong to me.

—Joseph A. Strapac

In the 1940s, eastward freight 824 was scheduled out of Los Angeles Yard at seven a.m. every day, destined for Colton. Beyond that point it became an extra train operating in CTC territory. We're treated to a worm's-eye view of #5007 powering 824, highlighting the third cylinder located directly below the smokebox—an attempt by Alco and SP to improve tractive effort, but in reality a maintenance headache. *(Southern Pacific Historical & Technical Society: Towler collection)*

Chapter One
Introduction

The Southern Pacific Company operated rail lines between Portland and New Orleans and from the San Francisco Bay Area to Ogden. A complex corporate history began in Texas, Louisiana and California in the 1850s, gradually being knitted together by the holding company incorporated in 1884. For operating purposes (as well as legal constraints), the railroad was operated as two separate primary systems, split at El Paso, Texas. West of El Paso, the railroad called itself the "Pacific Lines," using the corporate initials "S.P." East of El Paso, rail operations were considered the "Texas & Louisiana Lines," with a separate corporate identity of the Texas & New Orleans, using the initials "T. & N. O."

Both the SP and the T&NO expanded through new construction and via purchase of connecting rail lines. In some notable cases, it was expedient to set up separate rail operating organizations, often with ownership shared by other railroads or entrepreneurs. Many of these subsidiaries retained their identities into modern times, including the Southern Pacific of Mexico, Northwestern Pacific, Pacific Electric, and the San Diego & Arizona Eastern.

The Pacific Lines and its predecessors purchased over 2,700 steam locomotives between 1855 and 1956. Of these, only about 1,400 survived until 1940 while some 1,300 had been previously dropped as they became obsolete. The first great purge took place when Edward Harriman's Union Pacific took control of the SP in 1901; by 1904, UP's "Associated Lines" purchasing policies brought SP hundreds of new Common Standard locomotives—allowing the railroad to retire fleets of obsolete machines.

Locomotive purchasing responds to the nation's business cycle as well as local traffic fluctuations. In busy times SP needed all the locomotives it could get. On the other hand, railroad management has always been graded on "locomotive utilization:" the higher the percentage of owned locomotives actually in use, the better for all concerned. Business slowdowns meant putting locomotives—even new ones—into dead storage, because the traffic wasn't there to move. Operating crews and shop personnel are laid off and the railroad stops buying new locomotives.

In 1913, the courts forced the Union Pacific to give up its control of the Southern Pacific, but by then SP had accumulated a large fleet of locomotives that were to serve the railroad well for decades. The onset of World War I caused rail traffic to boom, but the locomotive builders were busy with military and foreign orders. A nationwide locomotive shortage resulted; in order to get the locomotives it needed, SP began building more Common Standard locomotives at company shops in Ogden, Los Angeles and Sacramento.

During World War I, SP was too busy to retire many locomotives. Once the war was over, however, it was time to retrench and modernize the roster. A program to replace old 0-6-0 switchers with new ones saw superheated machines built at company shops and the Lima Locomotive Works: forty 0-6-0s, #1255-1294, were added to the roster between 1920 and 1924.

SP was ready to take advantage of advances in locomotive technology. In 1921, Baldwin delivered completely new designs of 4-6-2 (#2461-2475) and 2-10-2 (#3652-3667) that made use of wagon-top boilers and Delta trailing trucks. Both the freight 2-10-2 and passenger 4-6-2 ordering programs were to continue into 1924.

Passenger trains were growing heavier—not just because of steel cars, but as a result of increases in head-end traffic. SP turned to Alco for its first 4-8-2s (#4300-4309) in 1923 and came back for eighteen more (#4310-4327) in 1924. SP accountants figured that the Sacramento Shops could build 4-8-2s significantly cheaper than Alco, if not at the same production rate. Management approved a 4-8-2 production line that would deliver forty-nine new 4-8-2s (#4328-4376) over the period 1925-1930.

In 1924, SP's own locomotive purchasing programs added ten 0-6-0s, eight 4-6-2s, eighteen 2-10-2s and eighteen 4-8-2s. At the same time, however, the Pacific Lines roster received a double dose of secondhand locomotives, as both the El Paso & Southwestern and the Arizona Eastern rosters were acquired. Twenty-four locomotives from the AE (all built to SP blueprints) and 136 "not invented here" coalburners from the EP&SW (in addition to the 54 new locomotives acquired that same year) joined the SP fleet in 1924—effectively halting SP's purchases of everything but 4-8-2s and 4-10-2s for the next three years.

The former EP&SW locomotives were at first confined to their original home territory until they could be modified to conform to SP's own shop practices. Part of the EP&SW lease contract required the use of coal fuel for locomotives in that territory, so SP eventually had to convert a few oil-burning 2-10-2 locomotives to accept coal fuel. In 1939, SP had Lima build a dozen new coal-fired 2-8-8-4s (#3800-3811).

After an expensive 1925-1927 venture into three-cylinder locomotion (Alco 4-10-2s #5000-5048), SP never again purchased an Alco steam locomotive. Advertised performance gains by these long, rigid wheelbase machines were outweighed by reduced availability and increased maintenance costs.

Compound steam as used in Mallet articulated locomotives had been touted as an economy measure, but by the mid-twenties, SP realized that while compound locomotives were too slow, their boilers had sufficient capacity to power four simple cylinders.

In June 1927, SP tested this thesis by converting 1911-vintage 2-8-8-2 compound #4041 to simple steam operation, with 22-inch diameter cylinders replacing the original 26-inch high pressure and 40-inch low pressure cylinders. Positive results from this project granted new life to these old cab-aheads; six more were converted to simple in 1928 and the resulting program eventually included almost all of them.

Lessons learned from the 2-8-8-2 program were applied

to a much more modern design of cab-ahead that more than doubled both the firebox heating surface and that of the boiler as well. It would take a four-wheel leading truck to bear the increased firebox weight, but the leap to a 63-inch driver 4-8-8-2 design finally brought SP the locomotive it needed: one that could get around mountain curves and yet deliver both speed and power on the flatlands.

Baldwin built the first 4-8-8-2 cab-aheads (#4100-4109) in September and October 1928. They were an instant success, to be repeated with fifteen more in 1929 and then twenty-five more (#4126-4150) in 1930. The railroad's first 4-8-4s arrived in 1930. Baldwin built ten (#4400-4409), conservatively developing the design around SP's successful Mt-5 4-8-2. They featured 10% greater heating surface, significantly greater steam pressure and slightly smaller cylinders than their predecessors.

In the early 1930s, a business depression wiped out the gains of the previous decade as long lines of stored locomotives rusted to the rails at every major terminal. The depression forced SP to cleanse its roster of hundreds of obsolete locomotives. Not until 1936-37 did business pick up sufficiently for the SP to begin buying new locomotives again.

From the locomotive historian's standpoint, between 1930 and 1940, the railroad was forced to clean house of most of the really old (and interesting) locomotives that had survived the Harriman years.

By 1936, the Pacific Lines locomotive roster was relatively lean. Railroad managers could see clouds of war gathering in Europe as early as 1938, at the same time that traffic began returning to normal levels. Remembering the lessons of 1914-1918, SP began to purchase modern steam locomotives in significant quantities: Lima 4-8-4s for new streamliners and Baldwin cab-aheads for freight service. War again broke out, causing SP to accelerate its locomotive purchasing program before locomotive production was diverted to war needs.

Six new streamlined Lima 4-8-4s based on the Baldwin GS-1s were built in late 1936 for the new *Daylight* train. They were followed in 1937 by fifteen improved Lima machines. On the freight side, Baldwin delivered twenty-five enhanced cab-aheads (class AC-7, #4151-4176) that would set the pattern for future steam freight locomotive purchases. No new locomotives were bought in 1938, but in 1939, AC-8s #4177-4204 and conventional cab AC-9s #3800-3811 were added to the roster. Lima supplied GS-4s #4430-4449 in 1941.

Faced with a brand-new locomotive technology, the diesel-electric freight locomotive, SP management chose not to risk changing (iron) horses in mid-stream. Rather than buy a new technology and its supporting facilities, SP opted to *not* get in line for diesel freight locomotives during the war years. Steam locomotives could be built without resorting to light-alloy steels, while diesel production was tightly allocated to serve the needs of both the military and the railroads.

In 1942, Lima delivered GS-4 and GS-5 locomotives #4450-4459, while Baldwin supplied fifty new cab-aheads (#4205-4254). Even though the government was severely rationing new locomotive construction, it did allow continuing locomotive production from existing blueprints. In 1943, cab-aheads #4255-4287 and 4-8-4s #4460-4469 were delivered. The GS-6 locomotives had to be ordered as replicas of the 1936-vintage GS-2s, without the streamlining but with the addition of vestibuled cabs. Finally, in 1944, Baldwin shipped its final seven 4-8-8-2s, class AC-12 #4288-4294.

The end of the war found the SP with a fleet of relatively modern—but worn—steam locomotives and a few dozen diesel switchers. Financially recovered from the depression, SP emerged from the war years with cash to finance a program of new locomotive purchases, but the locomotive builders were backlogged with orders. SP would have to live with delivery delays, which meant that its total dieselization program would have to take place over a period of years. New diesels would not begin to arrive until 1947.

In 1948, SP released its ten-year plan for dieselization. From the 1,541 steam locomotives then on the roster, in the next ten years it was planned to amortize and retire about 86% of the fleet—some 1,325 locomotives—retaining only 216 of the most modern machines. By 1958, the SP steam roster would be reduced to twelve 2-8-8-4s, 144 4-8-8-2s and 60 4-8-4s. This schedule assumed that diesels would take over the assignments then held by smaller, older steam locomotives and that SP could recover full depreciation of its modern steam locomotives before retiring them.

Management at that time had insufficient information regarding road diesel locomotive productivity; furthermore, they could not predict new locomotive models that would deliver even more performance and reliability. Within a few years it was realized the dieselization *by district* (and consequent retirement of obsolete steam maintenance facilities) was more cost-effective than indiscriminately retiring steam locomotives just because they were fully amortized or used only seasonally. Retaining small, well-maintained steam engines allowed SP to postpone total dieselization while buying new road diesels to take over regions where steam was expensive to operate.

Dieselization began in 1947 with the goal of eliminating large steam freight locomotives from bad-water districts, including the Salt Lake, Rio Grande and Tucson Divisions. Only a trickle of new diesel passenger units arrived each year, since SP had few streamlined passenger trains and an ample supply of steam passenger locomotives. Diesel switchers were also added each year, but they worked side-by-side with surviving steam switchers.

Deliveries of diesel units continued into the 1950s as steam locomotives were being disposed of. There was no particular "grand purge," but old steam locomotives were not being replaced by new ones. Some were sent to the T&NO to run out their final miles. In December 1951, the Southern Pacific of Mexico was sold to the Mexican government along with more than 90 serviceable steam locomotives.

SP was in no hurry to rid itself of its steam locomotives; an average of about 130 steam locomotives per year disappeared from the Pacific Lines roster each year between 1949 and 1958. Until 1951, most retired steam locomotives were stripped for parts and the remaining hulks cut up for scrap by SP employees. As shops were closed and the need diminished for spare parts, SP began selling locomotives for scrap. With water and fuel oil drained and reciprocating valve gear and piston rods torched to reduce need for rolling lubrication, retired steam locomotives were moved to scrap dealers in Los Angeles and the Bay Area.

The year 1956 brought an end to steam freight and switching operations on the Pacific Lines. In late November and early December of that year, SP ran its last steam freight runs. Three 2-8-0 locomotives (#3400, 3406 and 3423) remained available for service on the Nacozari Railroad out of Douglas, Arizona, while the 71-mile narrow gauge line held one 4-6-0 serviceable as protection power for its lone diesel. The Arizona steam operation quietly expired in 1959, while the entire narrow gauge was closed forever in 1960.

Many SP steam locomotives were donated for display; over fifty survive today—but most of them have been out in the weather for more than fifty years. Although a few SP locomotives once consigned to permanent display have been restored to FRA-approved rail passenger service, for practical purposes the age of steam on the Southern Pacific was closed for good in early 1957, when the last steam-powered Commute train tied up at San Jose.

ABOUT THIS BOOK:

Photographing Southern Pacific steam locomotives as a hobby has been practiced for more than a century. David Joslyn, an SP employee in Sacramento, took locomotive pictures for his own interest before his boss assigned him the job of Shops Photographer. In the Bay Area around 1904-06, Robert McFarland began taking pictures of SP steam with a roll-film camera. Other photographers and collectors began following the same steps.

By the late 1930s, SP steam locomotives had a large, well-organized following. Once prosperity returned after the Depression, young men in both northern and southern California began accumulating images with their own cameras or by trading negatives and prints with other collectors. In 1941, the Railway & Locomotive Historical Society in Boston took a big risk and published Gerald Best's softcover, all-time roster book on SP locomotives; it was a surprise hit.

After World War II, the cadre of steam fans anticipated dieselization and set out to capture images of every surviving steam locomotive. Guy Dunscomb, noting this interest, in 1949 self-published one of the earliest railfan pictorials, *Maps of the Southern Pacific*, basically an album of divisional maps but including thumbnail images of SP steam locomotives. In 1956, the R&LHS revised Best's 1941 work as its *Bulletin 94*, while Dunscomb produced his monumental *A Century of Southern Pacific Steam Locomotives* in 1962.

The book you are reading is patterned after Dunscomb's *Century*, but is geographically limited to just the Pacific Lines and in time scope to the World War II era and the dozen years afterward that marked the period when SP became fully dieselized. Its goal is to present quality images of Pacific Lines steam locomotives that were photographed during the 1940-1957 time period and to accompany those images with explanatory captions and tabular material.

The roster is necessarily condensed, but detailed, comprehensive roster information is available elsewhere, both in print and on the Internet. It is anticipated that the Society will release a number of pertinent SP documents in digital form in the near future. See the bibliography later in this chapter and consult the Society's Website at *www.sphts.org.*

One of the first photographers to exploit photo opportunities at Taylor Yard near Los Angeles, and certainly one of the most productive chroniclers of the Southern Pacific, Gerald M. Best contributed immensely to this book. His work set the standard for locomotive photography and he had open access while mechanics attending 4-10-2 #5021 must deal with a misbehaving firing valve. *(G. M. Best)*

Selected Bibliography

Best, Gerald M., *Iron Horses to Promontory.* (Golden West Books, 1969)

Boynton, James E., *Three Barrels of Steam.* (Glenwood Publishers, 1973)

Church, Robert J., *Cab-Forward.* (Signature Press, 1982)

Church, Robert J., *Southern Pacific's 4300 4-8-2s.* (Signature Press, 1996)

Church, Robert J., *SP Daylight Locomotives.* (Signature Press, 2004)

Demoro, Harre W., *Southern Pacific Bay Area Steam.* (Chatham Publishing Co., 1979)

Diebert, Timothy S. and Strapac, Joseph A. *Southern Pacific Steam Locomotive Compendium.* (Shade Tree Books, 1987)

Dunscomb, Guy L., *A Century of Southern Pacific Steam Locomotives.* (Various editions, 1962-)

Lohse, Richard and Bergen, Larry. "SP's Final Regular Steam Operations." *SPH&TS Trainline* #95 (Spring 2008)

McLennan, A. D. "Espee Mainline Steam Operations Yuma to El Paso." *SPH&TS Trainline* #81 (Fall 2004)

Menke, Arnold S., *The Compendium Companion.* (Privately published, February 2002)

Photo Credits

Tom Gildersleeve, on his way home during spring break at Stanford, took our front cover image on March 24, 1957 at Farwell in Niles Canyon on one of the last steam-powered excursions on the SP, the first after official dieselization in January 1957 and perhaps the next-to-last using a 4-8-4; the train was *en route* to Fresno via the West Valley Line. Tom's camera was an f:2.8 Stereo Realist with a top shutter speed of 1/200th of a second.

Dustjacket design by David Styffe.

Rear Cover

Rusting to the rails at Taylor Roundhouse in January 1949, one of the oldest steam locomotives still on the SP roster (dating from December 1900 at the Cooke Works in New Jersey) illustrates how quickly neglect catches up with unused steam locomotives. Fortunately, in July 1950, SP found an assignment for the #1645, leasing it to the San Diego & Arizona Eastern until December 1951, when it was brought back to Los Angeles and sold for scrap in February 1952. *(F. J. Peterson: Miller collection)*

Front Endpaper

The Southern Pacific never operated many deluxe trains, but in the 1930s and 1940s, SP offered luxury, scenery, and "the most beautiful train in the world," the *Daylight*, between San Francisco and Los Angeles. In their peak years, this corridor was served by a *Morning Daylight* and a *Noon Daylight* as well as the overnight, all-Pullman *Lark*, cosseting customers who wanted to ride in style. In 1938, soon after its debut, the *Noon Daylight* (train 96) passes Potrero Tower as it gathers speed departing San Francisco. GS-3 #4424 provides the power today. *(R. H. McFarland: Menke collection)*

Rear Endpaper

Even in the closing summer of steam freight train operations, the SP assembled interesting little trains for its Harriman-era 2-6-0 and 2-8-0 locomotives. For instance, 2-8-0 #2852 and wood caboose #4 join forces to deliver seven carloads of sugar beets through Newark in 1956. *(John Harder: Menke collection)*

Frontispiece

What would Lucius Beebe have to say about this scene? Perhaps "man and machine prepare to battle Beaumont" would do the trick. This undated photo finds Southern Pacific 4-10-2 #5012 ready to leave Colton for Indio, while the unidentified engineer waits for the highball. *(Thomas Hotchkiss)*

Title Page

Train 408 crests the summit of the Coast Range at Altamont with cab-ahead #4232 in charge. April 1954 was not quite the final season for big steam freight operation in the Bay Area, but SP had not purchased a new steam locomotive in ten years and new diesels were arriving daily. The end was in sight and the photographer can be forgiven for cutting all his morning college classes to record this moment. *(Charles Givens)*

SOUTHERN PACIFIC STEAM LOCOMOTIVES

Pacific Lines: 1940-1957 Era

Type	*Class*	*Numbers*	*Builder*	*Built*	*Arrived*	*Notes*
Narrow Gauge						
4-6-0	T	8	Baldwin	1907	9/29	Nevada-California-Oregon 8
	T	9	Baldwin	1909	9/29	Nevada-California-Oregon 9
	T	14	Baldwin	1886	7/06	South Pacific Coast 18
	T	17	Baldwin	1887	10/07	South Pacific Coast 21
	T	18	Baldwin	1911	9/29	Nevada-California-Oregon 12
	T	22	Schenectady	1899	9/29	Nevada-California-Oregon 22, F&CC 22
Shop Switch (Maintenance of Way number series)						
0-6-0T	S-5	208	Baldwin	1901	8/39	SP 1079; Brooklyn Roundhouse
	S-5	209	Baldwin	1902	12/35	SP 1097; West Oakland Roundhouse
0-6-0	S-2	211	Rhode Island	1888	12/35	SP 1015; Lucin Quarry
	S-1	212, 214	Schenectady	1888	12/35	SP 1024, 1029; Lucin Quarry
	S-3	216	Schenectady	1881	12/35	SP 1072; Lucin Quarry
0-6-0T	S-8	217	Alco-Brooks	1908	8/40	SP 1295; West Oakland Roundhouse
	S-8	2nd 220	Baldwin	1907	10/42	SP 1162; Sacramento General Shops
	S-3	221	Rogers	1881	1/15	SP 1070; Taylor Roundhouse
	S-1	564	Schenectady	1888	9/17	SP 1027; Tucson Roundhouse
	S-5	565	Baldwin	1903	8/39	SP 1109; Los Angeles General Shops
	S-5	2nd 567	Baldwin	1903	2/40	SP 1108; Taylor Round House (*sic*)
	S-5	2nd 568	Baldwin	1903	2/46	SP 1110; Sparks Shops
	S-8	2nd 569	Baldwin	1907	12/47	SP 1155; Ogden Shops
0-6-2T	S-1	571	Sacramento Shops	1882	6/16	SP 1901; Ogden Shops
0-6-0T	S-8	572	Baldwin	1911	8/48	SP 1191; Roseville Roundhouse
	S-1	578	Schenectady	1888	11/19	SP 1031; Los Angeles General Shops
	S-2	619	Rhode Island	1888	8/21	SP 1020; Roseville Roundhouse
	S-1	620	Schenectady	1893	12/28	SP 1041; El Paso General Shops
	S-17	966	Alco-Schenectady	1907	2/37	SP 1010; Bayshore Roundhouse
Standard Gauge, Revenue Service Locomotives						
0-6-0	S-22	2nd 1100	Alco-Brooks	1922	4/40	South San Francisco Belt 5 *(Saturated steam)*
	S-5	1110	Baldwin	1903		
	S-7	1112-1121	Alco-Schenectady	1904		*Saturated steam; slide valves*
	S-8	1122-1174	Baldwin	1906-07		
	S-8	1175-1178	Alco-Brooks	1908		
	S-8	1179-1194	Baldwin	1909, 1911		
	S-9	1195-1209	Baldwin	1912		Slide valve; 1209 rebuilt with piston valves
	S-10	1210-1221	Baldwin	1913		*Superheated; piston valves*
	S-11	1222-1231	Lima	1914		
	S-10	1232-1246	Baldwin	1918		
	S-12	1247-1284	Sacramento Shops	1919-23		
	S-14	1285-1294	Lima	1924		
	S-8	1295, 1297	Alco-Brooks	1908	11/24	Arizona Eastern 36, 38 *Saturated steam; slide valves*
	S-8	1296	Baldwin	1909	11/24	Arizona Eastern 37 *Saturated steam; slide valves*
	S-10	1298-1299	Baldwin	1917-18	11/24	Arizona Eastern 39-40 *Superheated; piston valves*
	S-10	2nd 1240	Baldwin	1918	11/53	T&NO 134
	S-13	2nd 1283	Baldwin	1920	9/52	T&NO 140
	S-13	2nd 1288	Baldwin	1921	11/53	T&NO 142
	S-14	2nd 1295	Houston Shops	1926	6/52	T&NO 165
	S-13	2nd 1296-97	Baldwin	1921	6/52	T&NO 144, 146
0-8-0	SE-2	1400-1401	Baldwin	1901, 1903	11/24	SP 1302, 1306; EP&SW 51, 55
	SE-4	1402-1409	Sacramento Shops	1930-31, 37		SP 1307-1314 *(Re# 7/48 to 4501-4508)*

Type	Class	Numbers	Builder	Built	Arrived	Notes
4-4-0	E-23	1500	Schenectady	1898		*Leased to NWP.*
	E-23	1502	Cooke	1900		*Stationary boiler at Indio*
2-6-0	M-4	1615-1618	Schenectady	1899		
	M-4	1619-1628	Cooke	1899		
	M-4	1629-1631	Schenectady	1900		
	M-4	1632-1649	Cooke	1901		
	M-4	1650-1651				*Numbers not used*
	M-4	1652-1672	Cooke	1899-1900		
	M-4	1673-1677	Schenectady	1900		
	M-4	1678-1707	Cooke	1900		
	M-4	1708-1709	Schenectady	1900		
	M-4	1710-1719	Cooke	1901		
	M-4	2nd 1660	Cooke	1899	11/24	Arizona Eastern 577
	M-4	2nd 1713	Cooke	1900	11/24	Arizona Eastern 575
	M-8	1720-1724	Baldwin	1901		
	M-6	1725-1769	Baldwin	1901		
	M-8	1770-1779	Baldwin	1902		
	M-6	1780-1803	Baldwin	1902-03		
	M-9	1804-1818	Alco-Brooks	1908		
	M-9	1819-1822	Baldwin	1909		
	M-6	1823-1825	Sacramento Shops	1917-18		
	M-9	1826-1827	Sacramento Shops	1918		
	M-9	1828-1830	Alco-Brooks	1908	11/24	Arizona Eastern 567-569
	M-11	1831-1836	Baldwin	1909	11/24	Arizona Eastern 570-574, 576
	M-21	1837-1845	Houston Shops	1928-30	1939-41	SP de M and T&NO 520-524, 526-529
4-6-0	T-36	2nd 2103	Alco-Manchester	1902	11/24	EP&SW 115
	T-37	2nd 2105-06	Alco-Rhode Island	1903	11/24	EP&SW 125-126
	T-9	2161, 2172	Schenectady	1888		
	T-8	2174, 2178	Schenectady	1888		
	T-1	2242~2271	Cooke	1895, 1897		
	T-26	2283~2299	Baldwin	1901-02		
	T-23	2301-2310	Baldwin	1903		
	T-28	2311-2330	Baldwin	1907		
	T-28	2331-2342	Alco-Brooks	1908		
	T-28	2343-2352	Baldwin	1909, 1911		
	T-31	2353-2362	Baldwin	1912		*Walschaerts valve gear*
	T-32	2363-2370	Alco-Brooks	1913		
	T-40	2371	Sacramento Shops	1917		
	T-32	2372	Sacramento Shops	1918		
	T-32	2373-2375	Los Angeles Shops	1918-19		
	T-32	2376-2384	Sacramento Shops	1919-20		
	T-57	2385	Baldwin	1906	6/41	San Diego & Arizona Eastern 20
	T-58	2386	Baldwin	1907	6/41	San Diego & Arizona Eastern 26
4-6-2	P-1	2400~2404	Alco-Schenectady	1904		
	P-1	2405~2427	Baldwin	1906-07		
	P-4	2401-2402	Los Angeles Shops	1928-29		*Rebuilt from P-1*
	P-4	2409-2410	Los Angeles Shops	1927-28		*Rebuilt from P-1*
	P-4	2414, 2419	Los Angeles Shops	1928		*Rebuilt from P-1*
	P-4	2420, 2424	Los Angeles Shops	1928-29		*Rebuilt from P-1*
	P-4	2422	Sacramento Shops	1927		*Rebuilt from P-1*
	P-4	2436	Los Angeles Shops	1928		*Rebuilt from P-3*
	P-3	2428~2437	Baldwin	1911		
	P-5	2438-2452	Baldwin	1912		*Walschaerts valve gear*
	P-6	2453-2458	Alco-Brooks	1913		*2455-57 to T&NO 650-652 in 1937*
	P-3	2459-2460	Sacramento Shops	1917-18		*Built from spare parts during WWI*
	P-8	2461-2475	Baldwin	1921		
	P-7	2476-2477	Lima	1917	2/22	Arizona Eastern 701-702
	P-10	2478-2491	Baldwin	1923-24		

Type	*Class*	*Numbers*	*Builder*	*Built*	*Arrived*	*Notes*
2-8-0	C-32	3rd 2500	Baldwin	1906	4/42	Ferrocarril de Nacozari 103
	C-15	2505-2507	Baldwin	1900	11/24	El Paso & Southwestern 181-183
	C-17	2510-2511	Baldwin	1902, 1905	11/24	El Paso & Southwestern 217-218
	C-9	2513-2561	Baldwin	1907		
	C-9	2562-2599	Baldwin	1906		
	C-2	2601-2602	Schenectady	1899		
	C-3	2608-2610	Schenectady	1900		
	C-4	2612~2623	Schenectady	1901		
	C-5	2624-2625	Baldwin	1901		
	C-5	2629-2630	Baldwin	1901		
	C-5	2632, 2635	Baldwin	1901		
	C-5	2640-2641	Baldwin	1901-02		
	C-5	2644~2667	Baldwin	1902		
	C-5	2669~2690	Baldwin	1903		
	C-8	2698-2725	Baldwin	1904		
	C-8	2727-2749	Alco-Schenectady	1904		
	C-10	2750	Alco-Schenectady	1904		
	C-8	2751	Alco-Schenectady	1904		
	C-9	2752-2795	Baldwin	1905		
	C-9	2796-2799	Baldwin	1906		
	C-9	2800-2821	Baldwin	1907		
	C-9	2822-2830	Alco-Brooks	1907-08		
	C-10	2831-2836	Baldwin	1911		
	C-10	3rd 2837-2838	Baldwin	1911	11/24	Arizona Eastern 669-670
	C-10	2839-2842	Sacramento Shops	1917-18		
	C-10	2843-2844	Los Angeles Shops	1917-18		
	C-10	2845	Sacramento Shops	1918		
	C-10	2846	Los Angeles Shops	1918		
	C-10	2847-2855	Sacramento Shops	1918-19		
	C-10	2856-2857	Ogden Shops	1919		
	C-9	2858-2860	Alco-Brooks	1908	11/24	Arizona Eastern 671, 672, 668
	C-11	2861-2864	Baldwin	1920-21	1/56	St. Louis Southwestern 751, 753, 755, 768
	C-12	2865-2868	Baldwin	1923	1/56	St. Louis Southwestern 773, 777, 781, 783
4-8-0	TW-1	2900~2913	Schenectady	1895		
	TW-8	2914-2923	Schenectady	1898		
	TW-4	2926~2931	Cooke	1882-83		
	TW-3	2932~2943	Schenectady	1891-92		
	TW-3	2945	Schenectady	1888		
	TW-2	2947~2952	Schenectady	1892		
	TW-6	2957	Schenectady	1893		
4-4-2	A-6	3000-3001	Sacramento Shops	1927		*Rebuilt from A-3*
	A-6	3002-3003	Los Angeles Shops	1928		*Rebuilt from A-3*
	A-3	3025, 3029	Alco-Schenectady	1904		
	A-3	3032, 3036	Alco-Schenectady	1904		
	A-3	3052	Baldwin	1906		
	A-3	3057, 3060	Baldwin	1907		
	A-3	3071	Alco-Brooks	1908	2/13	SP de Mexico 851
4-6-2	P-11	3100~3109	Baldwin	1907	11/24	EP&SW 140~149
	P-12	3120-3129	Alco-Brooks	1917	11/24	EP&SW 160-169

Type	Class	Numbers	Builder	Built	Arrived	Notes
2-8-2	Mk-2	3200~3215	Baldwin	1911		
	Mk-4	3203, 3205	Ogden Shops	1928, 1931		*Rebuilt from Mk-2*
	Mk-4	3208, 3215	Shops	1929		*Rebuilt from Mk-2*
	Mk-4	3216-3235	Baldwin	1913		
	Mk-4	2nd 3236-3240	Baldwin	1913	11/24	Arizona Eastern 901-905
	Mk-5	3241-3249	Baldwin	1913		
	Mk-6	3250-3269	Lima	1914		
	Mk-5	3270	Baldwin	1914		
	Mk-5	3271-3275	Baldwin	1913		*Renumbered from 1st 3236-3240*
	Mk-5	3276-3277	Lima	1917	11/24	Arizona Eastern 906-907
	Mk-10	3295-3296	Alco-Brooks	1923	1/35	Minarets & Western 102, 104
	Mk-11	3297-3298	Alco-Brooks	1914	11/40	Newaukum Valley 521-522
	Mk-7	3300-3309	Alco-Schenectady	1913, 1916	11/24	EP&SW 363-372
	Mk-8	3310-3314	Alco-Schenectady	1918	11/24	EP&SW 373-377
	Mk-9	3315-3324	Alco-Schenectady	1920	11/24	EP&SW 378-387
2-8-0	C-18	3400-3403	Baldwin	1902	11/24	EP&SW 251-254
	C-18	3406	Baldwin	1902	11/24	EP&SW 257
	C-18	3408-3409	Baldwin	1902	11/24	EP&SW 259-260
	C-19	3410-3411	Baldwin	1904	11/24	EP&SW 261-262
	C-19	3415, 3418	Baldwin	1904	11/24	EP&SW 266, 269
	C-19	3420-3421	Baldwin	1904	11/24	EP&SW 271-272
	C-19	3423-3426	Baldwin	1904	11/24	EP&SW 274-278
	C-26	3440-3444	Baldwin	1906	11/24	EP&SW 280-284
	C-27	3445-3449	Baldwin	1907	11/24	EP&SW 285-289
	C-29	3450-3451	Alco-Schenectady	1907	11/24	EP&SW 290-291
	C-28	3452-3469	Alco-Schenectady	1907	11/24	EP&SW 292-309
2-8-4	B-1	3500-3509	Lima	1928	7/45	Boston & Maine 4000~4019
2-10-2	F-1	3611~3651	Baldwin	1919		*Many sold to T&NO*
	F-1	3652	Alco-Brooks	1918	1/21	Arizona Eastern 1001
	F-3	3653-3667	Baldwin	1921		
	F-4	3668-3717	Baldwin	1922		
	F-5	3718~3769	Baldwin	1923-24		*Many sold to T&NO*
	F-5	3770	Baldwin	1923	3/53	T&NO 910; originally SP 3738
2-8-8-4	AC-9	3800-3811	Lima	1939		
4-6-6-2	AM-2	3900-3911	Baldwin	1911		
2-6-6-2	MM-3	3930-3931	Alco-Schenectady	1920	2/43	Verde Tunnel & Smelter 500-501

Soon to be on its way with train number 783, westward toward Fresno, class F-1 2-10-2 #3613 (built by Baldwin in 1919) pumps up the train air prior to departure from Bakersfield on October 11, 1952. While banished from Tehachapi and the Sierra, ten-driver machines such as this were still useful in the early 1950s on shuttle trains between Bakersfield, Fresno, Tracy, Oakland and Roseville.
(Allan S. Youell)

Type	Class	Numbers	Builder	Built	Arrived	Notes
2-8-8-2	AC-1	4000~4016	Baldwin	1909		
	AC-2	4017~4028	Baldwin	1911		
	AC-3	4029-4048	Baldwin	1912-13		
4-8-8-2	AC-4	4100-4109	Baldwin	1928		
	AC-5	4110-4125	Baldwin	1929		
	AC-6	4126-4150	Baldwin	1930		
	AC-7	4151-4176	Baldwin	1937		
	AC-8	4177-4204	Baldwin	1939		
	AC-10	4205-4244	Baldwin	1942		
	AC-11	4245-4274	Baldwin	1942-43		
	AC-12	4275-4294	Baldwin	1943-44		
4-8-2	Mt-1	4300-4327	Alco-Schenectady	1923-24		
	Mt-3	4328-4345	Sacramento Shops	1925-26		
	Mt-4	4346-4366	Sacramento Shops	1926-29		
	Mt-5	4367-4376	Sacramento Shops	1929-30		
	Mt-2	4385-4390	Alco-Brooks	1924	11/24	El Paso & Southwestern 410-415
4-8-4	GS-1	4400-4409	Baldwin	1930		
	GS-2	4410-4415	Lima	1937		
	GS-3	4416-4429	Lima	1937		
	GS-4	4430-4457	Lima	1941-42		
	GS-5	4458-4459	Lima	1942		*Roller bearings*
	GS-6	4460-4469	Lima	1943		
	GS-1	4470-4473	Baldwin	1930	1953	*Transferred from T&NO*
	GS-7	4475-4481	Baldwin	1930	6/53	St. Louis Southwestern 802~809
	GS-8	4485-4487	Pine Bluff Shops	1942	6/53	St. Louis Southwestern 815, 817, 818
	GS-8	4488	Pine Bluff Shops	1937	2/56	St. Louis Southwestern 813
0-8-0	SE-2	4500	Baldwin	1901	11/24	SP 1400, SPdeM 30, EP&SW 51, etc.
	SE-4	4501-4508	Sacramento Shops	1930-31, 37	7/48	Renumbered from 1402-1409
	SE-3	4509-4510	Houston Shops	1930, 1936	7/52	T&NO 855, 860
	SE-3	4511-4512	Houston Shops	1930	10/53	T&NO 851, 854
4-10-2	SP-1	5000-5015	Alco-Schenectady	1925		*Three cylinders*
	SP-2	5016-5038	Alco-Schenectady	1926		*Three cylinders*
	SP-3	5039-5048	Alco-Schenectady	1927		*Three cylinders*

Long-wheelbase 4-10-2 locomotives were found unsuitable for use over the Sierra; much of SP's 48-engine fleet (in the 5000 series) spent decades working eastward from Los Angeles. Grades here were just as demanding, but curves were more gentle than those between Roseville and Sparks. Eastbound between Garnet and Beaumont, class SP-3 #5045 leads a "reefer block" on August 27, 1952. In the far distance, the helper engine makes life a little easier for the road engine's crew. *(Southern Pacific Historical & Technical Society: Don Ball collection)*

The Pacific Lines' final *new* steam switchers were eight class SE-4 0-8-0s built at Sacramento Shops in 1930-37, using reclaimed boilers from retired 4-4-2 locomotives and new cast frames. Examples could be seen at Tucson, Los Angeles, Tracy and Roseville, but all of them retreated to northern California by late 1955. On the morning of November 6, 1954, 0-8-0 #4506 delivers a transfer cut across the Los Angeles River at Mission Tower. *(John E. Shaw, Jr.)*

Chapter Two
Switchers: 0-6-0 and 0-8-0

Providing switching service was one of the most expensive cost centers in railroading seventy years ago. Today we think in terms of truck or container pick-up and delivery, but in 1940 most freight moved in boxcars to customers' warehouses or factory doors, requiring plenty of switch engines at work all the time, moving those boxcars to their ultimate locations. SP addressed this question with a fleet of modern steam switchers—most of them 0-6-0s.

Eighty-five superheated, piston-valve 0-6-0s, #1210-1294, were built by the Company Shops, Baldwin and Lima between 1913 and 1924, constituting the most modern fleet of switch engines operated by any railroad in the far West. (Still more similar machines were purchased for the Texas & New Orleans.)

While most of the early Common Standard switchers (class S-7, #1112-1127, and class S-8, #1122-1194) remained on the roster at least into the late 1940s, these machines used saturated steam and slide valves and rolled on 51-inch drivers. In terms of modern technology, they were neither fast nor very productive. Beginning in 1913 with class S-10, however, SP's basic switcher design was refined with superheat, piston valves and—even later—with larger, 57-inch drivers to allow a turn of speed. While the S-10 through S-14 class 0-6-0s owed much to earlier Common Standard designs, SP took the concept to a high point with tender configurations that did not block the engine crew's rearward view.

By the time World War I-era locomotive shortages prompted SP to build its own, the class S-12 (thirty-eight of which were erected at the Los Angeles and Sacramento Shops between 1919 and 1923) had evolved to 200 pounds' boiler pressure and 57-inch drivers. Ten virtually identical S-14 locomotives (#1285-1294) came from Lima in 1924. They weighed about 77 tons in working order and generated 31,020 pounds of tractive effort; SP would utilize them intensively right up to the very end of steam operation, then donate many for permanent display.

The 0-6-0 switchers were used almost everywhere on the Pacific Lines, deployed to outlying locations as well as assigned to large yards and passenger terminals (where a favored few received special paint and plated metal trim and lettering). A few of the non-superheated 0-6-0s were rebuilt with saddle tanks and earned a new lease on life as shop switchers; they take a final bow in Chapter 16.

Eight-wheel switchers were uncommon on the SP. Seven Baldwin 0-8-0s, converted from road engines, were inherited from the El Paso & Southwestern in 1924 and assigned SP #s 1300-1306. In 1930, SP decided to reclaim boilers from eight class A-3 4-4-2 locomotives and mount them on new 0-8-0 chassis; these became class SE-4 #1307-1314, emerging from the Sacramento Shops between 1930 and 1937. In 1938, they were renumbered to #1402-1409 to clear their original numbers

(Continued on page 25)

The SP's modern 0-6-0 fleet—the most up-to-date in the western states—was typified by class S-12 and S-14 machines added to the roster between 1919 and 1924. Thirty-eight homebuilt S-12s (#1247-1284; erected at Sacramento and Los Angeles shops between 1919 and 1923) and ten Lima S-14s (#1285-1294) worked high-profile assignments until the very end of steam operations. Clean #1251 was working the Oakland Mole on February 6, 1955. *(Ray Whitaker)*

This 0-6-0 is actually *second* #1100, purchased in 1940 from the South San Francisco Belt, where it was #5. This 1922 Alco was similar to (but slightly larger than) SP class S-9, as well as the 0-6-0s of the nearby State Belt in San Francisco. Neither switching line adopted superheated steam and piston valves, although both selected modern valve gear for their 1920s-era 0-6-0s. SP assigned SB #5 class S-22 and gave it the then-vacant number 1100 that it would carry for the next eleven years. It is pictured in August 1940 at Bayshore, only four months after receiving SP paint and lettering for the first time. *(G. M. Best)*

Pre-Harriman era 0-6-0s in classes S-5 and S-6 (Baldwin 1901-1903) could not compare to the more powerful Common Standard locomotives that flooded the SP roster from 1904 to 1913. The only class S-5 0-6-0 locomotives to operate into the 1940s were shop switchers and the #1110, somehow missed in the scrap drives of the 1930s. Pictured in a "stored unserviceable" condition at Fresno on March 29, 1936, the #1110 would be restored to service and would work through World War II. In February 1946 it was rebuilt as shop switcher #MW568 for Sparks Shops, operating there until 1956.

The final refinement of SP's own design of 0-6-0s prior to Harriman Common Standards was class S-7, ten (#1112-1121) built by Alco-Schenectady in July 1904. They were never rebuilt with superheat, but did provide numerous design cues for subsequent S-8s. Pictured in 1950 at Taylor Roundhouse north of Los Angeles, #1118 would become the last class S-7 to be vacated (in June 1952) and was slated to become the El Paso shop switcher, but dieselization of the Rio Grande Division made that project redundant. The #1118 was never rebuilt and would be scrapped at El Paso in 1953.

Harriman management allocated seventy-three class S-8 locomotives to SP's Pacific Lines; they were built by Baldwin 1906-11 and Alco-Brooks in 1908 and numbered 1122-1194. The S-8s were postwar casualties of dieselization, rapidly vacated from the roster between 1946 and 1950. For instance, the #1152 was shopped in 1946, but by late 1947 (when this picture was taken) it was out of work and would be vacated in November and scrapped at the Brooklyn roundhouse that December. *(Don H. Roberts)*

Baldwin S-8 #1154 was photographed in May 1948 at Phoenix. By late 1951, almost every S-8 had already been scrapped or was stored unserviceable; only six survived to be scrapped in 1952. The exceptions were #1155, 1162 and 1191, rebuilt to tank engines for shop switcher service and #1171, retired in July 1949 to stationary boiler service at Bayshore, where it lasted until 1953. Transferred in the early 1950s to the Western Division as the Tucson Division dieselized its switching assignments, S-8 #1154 would be vacated in July 1951 at West Oakland and subsequently sold for scrap. *(E. B. Kille)*

Typical of SP's class S-8—even though the majority of them were erected by Baldwin—Alco #1178 is pictured on the "whiskers" tracks at Brooklyn roundhouse (across the Willamette River from downtown Portland) on May 15, 1947. Almost forty years old and still good for another three years, #1178 emerged from Alco's Brooks Works in March 1908 with an oval "Common Standard" builder's plate in the style dictated by the Harriman Associated Lines. It will be vacated November 24, 1950 and scrapped January 10, 1951 at Brooklyn.

Baldwin-built class S-9, delivered in 1912 as #1195-1209, introduced Walschaert valve gear for SP switchers. Other than that external feature and a boost in boiler pressure from 175 to 190 pounds, this class replicated the S-8 in almost every detail—retaining saturated steam and slide valves. The T&NO modified many of their S-9s with superheat and piston valves, but on the Pacific Lines, only #1209 was modified with piston valves but no superheater. At rest on the ready track in 1950 at the Mission Bay Roundhouse, the #1201 awaits its next assignment.

Havenscourt, in southeast Oakland, was located at approximately milepost 12.5. Passenger service at this location had been provided by SP subsidiary Interurban Electric Railway, but SP chose to continue freight locals and switching with steam and diesel power, rather than electrify sidings and yards and invest in electric switchers. Automobile cars for the nearby Chevrolet plant occupy 0-6-0 #1210 at Havenscourt in February 1941. *(W. C. Whittaker: Menke collection)*

Superheated steam with dry pipes and piston valves first appeared on SP switchers with the S-10 order from Baldwin (#1210-1221) in May 1913. While retaining its obsolete second cab-side windows, the #1215 (pictured in San Francisco in June 1956), has been modified with a power reverse installed above the running board, rather than the more common underslung arrangement. A narrow tender oil bunker allows better rearward vision at the expense of some fuel capacity. By this late date, steam switcher assignments were overshadowed by deliveries of new Fairbanks-Morse diesels. *(Herman Friedrich)*

After 1954, SP painters were instructed to letter only the class and weight on drivers on cab sides; furthermore, tender lettering began to show signs of overspray as steam maintenance budgets were cut. This mid-fifties image of the stored-serviceable S-11 #1229 shows it in good shape at West Oakland—but its chances of running again were slim. In late 1957, Lima #1229 would be formally vacated and donated the next year to the city of Roseburg, Oregon.

Thirty-eight class S-12 locomotives were built at the Sacramento and Los Angeles Shops between 1919 and 1923; while not as powerful as an 0-8-0, SP had an obvious success with this superheated 76-ton design and found it worthwhile to manufacture in-house, rather than buying commercially. A narrowed tender oil bunker improves sight lines to the rear. First of its class, the #1247 is pictured in its final assignment at Bakersfield on October 18, 1953. It will be vacated from the roster in 1956. *(Stan Kistler)*

An uncommon look at the back of a switcher cab emphasizes the good rearward visibility provided by the small "sausage" tenders (this one a class 47-C-2 built by Sacramento General Shops), while at the same time revealing just how open to the elements a steam locomotive cab could be. S-12 #1261 is working on the Wingfoot Spur, south of Los Angeles, on September 11, 1953. This trackage, serving a Goodyear Tire & Rubber plant, was switched by the Pacific Electric and the Santa Fe. *(Ray Ballash)*

Sacramento-built #1268 drills Commute coaches on an overcast June 18, 1956, as it assembles outbound trains for the evening rush at San Francisco. In earlier times, depot switchers were more or less permanently assigned and often given decorative paint jobs—but this occasion is the final summer of steam operation; in three months, S-12 #1268 will be vacated from the roster. The #1268 is clean, but not ornamented. *(James J. Buckley: Alan Miller collection)*

SP's Mission Bay Yard in San Francisco, less than a mile southeast of the passenger station at Third and Townsend, was a confined, unimpressive facility compared to sprawling Roseville or Taylor Yard in Los Angeles. Time is running out for twenty-nine-year-old Lima S-14 #1285; new diesel switchers are on order and the #1285 will be vacated in September 1955. San Francisco freight traffic was overwhelmingly inbound; wood PFE refrigerator cars and boxcars dominate this image, taken in April 1953.

(Arnold Menke collection)

The S-14 class became the final iteration of Espee's development of the six-wheel switcher—basically class S-12 as rendered by a commercial locomotive builder. In March and April 1924, Lima built ten each for the SP and the T&NO as SP #1285-1294 and T&NO #147-156. In this 1952 photo of the #1288, Lima's diamond-shaped builder's plate is still prominent on the smokebox. The #1290 was the first to leave the roster, in November 1952, followed in May 1953 by #1288, scrapped by the SP at Bayshore. The number 1288 would remain vacant for only a few months.

In 1952 and 1953, six recently shopped T&NO 0-6-0 switchers (no longer needed on home rails) were transferred to the Pacific Lines to run out their last miles; they were given second numbers in the SP's 1200 series: #1240, 1283, 1288 and 1295-1297. Class S-13 *second* 1288 was built by Baldwin in 1921 as T&NO #143, assuming its SP number in November 1953—just six months after SP had scrapped the newer *first* 1288, leaving that number slot open. Parked at West Oakland in May 1954, second #1288 will be scrapped in another ten months.

Even though the Southern Pacific constructed a new mainline southeastward along the Bay Shore out of San Francisco, the former main trackage was retained to serve as switching leads for the many industries and warehouses that still remained in the south-of-Market area. One of the most complex intersections was located at 10th and Division Streets (almost directly under today's 101-80 interchange), where SP rails shared the pavement with streetcar trackage—and Western Pacific switched a block to the east. Gleaming S-9 #1207 is delivering a boxcar. *(Brent MacGregor collection)*

for a planned order of diesel switchers, then renumbered again in 1948 to #4501-4508. They were used primarily at Roseville, Tracy and Los Angeles.

Four class SE-3 0-8-0 locomotives migrated west from the T&NO in 1952-53. They were home-built at the Houston shops with boilers reclaimed from retired 2-6-6-2 locomotives. Slightly heavier and more powerful than the SP version, they were numbered #4509-4512, serving only a few years in California before retirement.

With an oversupply of originating and terminating yards, the Southern Pacific was never truly able to escape its need for many switchers. Even into the diesel era, small yards and "peddler" operation demanded operation of small locomotives. Only in its very last years did the SP discover a surplus of switchers as yards were closed and more and more freight moved "overhead" via containerization and unit trains.

Class S-10 locomotives #1298-1299 came from subsidiary Arizona Eastern when that road was merged into the SP in 1924. The #1298, built by Baldwin in 1917, was originally AE #39. It is pictured on February 1, 1957 at West Oakland, where it had been shopped and placed into storage, anticipating future use in the summer rush of 1957 and perhaps even 1958. A business downturn changed SP's plans and the #1298 joined the list of steam engines slated for donation to on-line cities. As fate would have it, the #1298 would be the last steam locomotive to leave SP property, donated to Santa Cruz in March 1963. *(D. S. Richter collection)*

0-8-0 Locomotives

Built by Baldwin as a 2-8-0 in 1903, this chunky switcher worked for the El Paso & Southwestern before being sent to Mexico to spend twenty years as part of the Southern Pacific of Mexico roster. While assigned SPdeM #31, it was apparently never physically renumbered. In April 1947 it was declared surplus and moved dead-in-train to Los Angeles for scrapping. By the time photographers became aware of it, SP painters had blacked out the name on the tender; some intense "lobbying" took place and the road name was painted back on—where it remained until the railfans had their images! That truly brief moment in time took place on the Los Angeles dead line on May 14, 1947. *(G. M. Best)*

SP's class SE-4 locomotives, built as #1307-1314 at Sacramento between May 1930 and April 1937, were erected on new cast frames and boilers salvaged from retired class A-3 4-4-2 locomotives. SP made absolutely no changes to the old boilers, but did install new sport cabs. They were renumbered in March 1938 to #1402-1409, in anticipation of delivery of diesel switchers that did not take place. Their 1400-series numbers lasted for a little more than ten years, when they had to be renumbered again to the 4500-series to make room for diesels. This Los Angeles image dates from November 1939.
(Leo Munson collection)

SP 0-8-0 #4502 was erected at Sacramento in June 1930 as #1308, applying the boiler from 4-4-2 #3069 to a new cast frame. It was renumbered in March 1938 to #1403, and then renumbered again in July 1948 to #4502. A longtime denizen of the Roseville Yard facilities, it was pictured there on August 21, 1949. It would serve until retirement in the fall of 1956.
(Herman Friedrich)

SP had relatively little need for 0-8-0s, with modern 0-6-0s available and fleets of Harriman 2-8-0s always on hand. Eight 0-8-0s were built, eventually numbered 4501-4508, but their use was generally limited to the large yards at Roseville, Tracy and Los Angeles. Delivery stretched through the early 1930s; by the time SP was ready to buy more switchers, the diesel era was approaching and no more 0-8-0s were built. The #4501 was photographed on June 20, 1953, at Roseville.

The Texas & New Orleans built a dozen 0-8-0s (#175-186) at Houston between March 1930 and May 1936. New cast frames accepted shortened boilers from 2-6-6-2s; fireboxes remained the same, but #181-186 received sport cabs. In 1950-51, they were renumbered to #850-861, but were supplanted by diesels in a few years. Four, the #855, 860, 851 and 854, were moved to the Pacific Lines in 1952 and 1953, becoming SP #4509-4512. In late 1953, a freshly painted #4511 is pictured at Indio in transit to northern California.

As was the case with six T&NO 0-6-0s, the SP "bought" four 0-8-0s that still had ample flue time remaining, avoiding for a few years the purchase of four additional diesel switchers. Four SE-3 locomotives (all slightly heavier and more powerful than SP's home-built variety) became SP #4509-4512 in 1952 and 1953. They joined the SE-4s at Tracy and Roseville, but were set aside in fall 1956. The #4509, previously T&NO #855 and originally T&NO #180, is pictured in May 1955 at Tracy.

Diesel switchers began invading Taylor Roundhouse in Los Angeles after 1941. In 1947, SP constructed an elaborate "diesel house" a few steps away, but this facility was intended to serve road units. Switchers and roadswitchers shared the "whiskers" tracks with older steam engines; employees serviced both as necessary until steam locomotives (and their support tools) finally disappeared in 1956. In the early 1950s, a hood from a Baldwin roadswitcher and a Baldwin 2-6-0 both occupy one of the paint tracks at Taylor. *(Richard Steinheimer)*

Chapter Three
Moguls: 2-6-0

The Southern Pacific's 2-6-0 fleet was dominated by two large classes, the M-4s and M-6s; in fact, the bulk of the roster was delivered between 1899 and 1909, filling numbers 1615 to 1822. While very definitely obsolete almost from the moment of their arrival, the railroad found ample use for its 2-6-0s, especially in California.

In early 1899, the Huntington administration ordered more than a hundred locomotives of what would become class M-4 from both Cooke and Schenectady. Fifty were assigned to the Texas & Louisiana Lines, ultimately becoming their #410-459. Simultaneously, the Pacific Lines received #1440-1498 and #2100-2145—which were almost instantly renumbered #1615-1649 and #1653-1719. (Missing #1650-1651 were diverted at the factory to the Texas operation.) They were all simple, 73-ton coalburners as built. About two dozen were rebuilt in the 1920s with superheat, but most of the rest were scrapped in the early 1930s.

Fifteen M-8s (#1720-1724 and 1770-1779) and 72 class M-6 engines, all Baldwin Vauclain compounds, were built for the SP (#1725-1769 and 1780-1803) and the T&NO (#482-484) between 1901 and 1903. While paid for with Union Pacific money, they were designed to pre-Common Standard blueprints. Once converted to simple in the 1907-1913 era and then superheated in the years 1918-1923, these machines settled in for long, productive careers—many serving until the final days of SP steam.

Not until 1908-09 did the SP choose to buy true Common Standard 2-6-0s, ordering 15 class M-9 engines for 1908 delivery from Alco-Brooks and four more from Baldwin in 1909. These occupied #1804-1822, but others were allocated to the Southern Pacific of Mexico and to the Texas Lines.

Faced with a desperate locomotive shortage during World War I, SP gathered spare parts and was able to erect three M-6s (#1823-1825) and two M-9s (#1826-1827) at Sacramento in 1917-1918. In 1924, absorption of the Arizona Eastern added M-9s #1828-1830 and M-11s #1831-1836. Finally, from 1939 to 1942, nine heavy 2-6-0s from the T&NO were put to work on the Pacific Lines as #1837-1845, but they were not a success; SP supplanted them with 1000-horsepower Alco diesel switchers.

Of the twenty-three M-4 locomotives that survived past 1940, only four (#1617, 1627, 1632 and 1703) missed the superheating rebuilding program. All later 2-6-0s on the Pacific Lines were rebuilt with superheat, but none were considered sufficiently valuable to be rebuilt with modern valve gear.

Four M-6s and M-8s (#1720, 1745, 1750 and 1789) were scrapped in the 1930s. SP sold seventeen to the Ferrocarril del Pacifico in December 1951; they became FdelP #410-426. Almost all the rest remained on the active roster until the mid-1950s, on light road jobs and yard switching until SP was ready to dieselize these services. Likewise, the M-9 and M-11 locomotives served long and well.

A hundred twenty-nine identical class "ED" locomotives were built in 1899-1901 for the Pacific Lines and the T&NO, the last big "Huntington Standard" locomotive order before the Harriman era. They became class M-4 in 1913, but by then they were eclipsed by larger locomotives. Between 1919 and 1929, SP rebuilt 23 with superheat and piston valves; of these, sixteen survived into the 1950s. Retirement looms in two months for #1623 at El Paso on November 9, 1952. *(Fred A. Stindt)*

This January 1936 photo illustrates superheated M-4 #1644 on a short local at Glendale. By the postwar years, SP had an ample supply of larger surplus tenders and could scrap this little rectangular car. Another modernization was replacing this large headlight with a modern Pyle unit. Finally, reflecting the reality that SP had less use for small engines on road jobs, a switcher pilot would supplant this standard road engine affair. Stenciled on the pilot beam near the coupler lift lever is the #1644's divisional assignment and home roundhouse, in the form of a fraction "LA" over "LA." *(G. M. Best)*

After the war was over, #1644 acquired its modern look. A whaleback tender, new headlight and a new pilot are complemented by small "Southern Pacific" tender lettering and an aluminum-painted smokebox door. This 1901 Cooke product would never receive any further modernization and will be retired in July 1951.

Schenectady-built M-4 #1677 pauses for a portrait on July 11, 1940, at Modesto. Sacramento Shops added superheat and piston valves in August 1925, ensuring #1677's longevity after most of this class had been scrapped. After a brief postwar lease to the SP de Mexico, #1677 would be scrapped at West Oakland in June 1952. *(Douglas S. Richter)*

Every M-4 was built with saturated steam and slide valves; between 1919 and 1929, Pacific Lines shops rebuilt 23 of them with superheat and new cylinder/valve castings featuring piston valves and dry pipes. However, the T&NO did not have the budget to perform the same modification on all of its M-4s, frequently opting instead to utilize the less-expensive Cotter piston valve conversion that retained the expensive, heavy original cylinders and bolting on a casting carrying piston valves and dry pipe connections. One Cotter-converted M-4 (#433) went from the T&NO to the AE, becoming SP 2nd #1713 in 1924; it is pictured at Los Angeles on December 8, 1946. *(H. L. Kelso)*

One of the few saturated-steam, slide-valve M-4 locomotives to survive the Great Depression, the #1632 was pictured on February 19, 1938 on the Taylor Roundhouse lead track in Los Angeles. While completely intact, this 1900 Cooke product is actually in "stored serviceable" status, indicated by the canvas bagging on the turbogenerator. It will be scrapped in June 1940, before SP managers realized that they would need every locomotive they could find.

Unlike the #1632 above, M-4 #1685 was modernized in July 1922 at the Sacramento Shops with superheat and piston valves. Still useful in the late 1940s, it worked the postwar years on light locals in the Sacramento Valley. Even though it was given a large tender (as well as postwar paint), the #1685 carried this obsolete large headlight well into its sunset years. M-4 #1685 would be vacated in September 1953 and scrapped in November 1953 at the Sacramento Shops.

Built by Baldwin in 1901 as a Vauclain (four cylinder) compound, SP M-6 #1725 was rebuilt in October 1907 as a simple engine; superheating would be added in June 1921. Laid aside at Los Angeles in one of SP's enormous Depression-era dead lines, M-6 #1725 had not yet received a modern headlight when World War II called it back to active service—and then there seemed to be no time for cosmetic adjustments. Late in the War, in January 1945 at Los Angeles, this survivor retains its 1920s headlight location, but soon practical railroaders will bring it into conformance with the rest of the fleet. It will serve in southern California until retirement in 1952.
(H. L. Kelso)

One of the final redoubts for small steam engines on road jobs was the San Joaquin Division, which dispatched well-maintained 2-6-0s out into the foothills on branches out of Fresno and Bakersfield. Photographers took advantage of these priceless final moments, such as this occasion with the *Friant Local* at Pinedale on March 15, 1956, returning to Fresno. M-8 #1777 performs the honors with a well-timed soot plume. *(Stan Kistler)*

M-6 locomotives #1725-1769 were built by Baldwin in 1901. They were all delivered as four-cylinder Vauclain compounds, but SP rebuilt them with simple cylinders between 1907 and 1914 and then added superheat and dry pipes starting in 1918. They weighed ten tons more than the M-4s, but benefited from more extensive rebuilding and emerged as more capable machines. All but two worked into the 1950s; many were sold in 1951 when the SPdeM was sold to the Mexican government. The #1730 was photographed on October 11, 1953 at Los Angeles; it would operate until the end of SP steam operations. *(Fred A. Stindt)*

At Friant, northwest of Fresno in the foothills of the Sierra in the twilight of branch line steam operation, M-6 #1726 has picked up a small trainload of gravel, taken a spin on the Armstrong turntable and had its tender filled with water from the ample local supply—that's the Friant Dam in the right background. By the mid-fifties, small steam locomotives assigned to the Fresno roundhouse were treated as pampered pets, trimmed in an excess of aluminum paint. *(John E. Shaw, Jr.)*

During World War II, the Los Angeles area was so congested that SP diverted much of its drag freight over the parallel Pacific Electric line between Los Angeles and Colton. The PE did not have enough locomotives to shoulder the load, so SP steam power was routinely added to PE freights—always doubleheaded with a PE electric locomotive, both for additional power and to activate PE's overhead-triggered crossing signals. This eastbound is at LaVerne, near Pomona, in 1944 as SP M-8 #1721 leads PE motors #1630 and 1627.
(Craig Rasmussen collection)

In the depths of World War II, at Tracy on July 1, 1943, M-8 #1776 (known to crews as the "Liberty Bell") profiles its backup tender headlight, graphited smokebox and a light dispersion shield on the headlight. The M-8s were virtually indistinguishable from the more numerous M-6 engines; both classes offered useful performance on the flat profile of the San Joaquin Valley.
(David J. Welch)

Moguls were used up to the end of steam operations in both the San Joaquin and Imperial Valleys. They weren't tremendously powerful, but they were relatively fast—and could be trusted to deliver perishables or livestock without the crew "going dead on the law." M-6 #1768 has been downgraded from road service with utilitarian switcher footboards, but sports a good-sized Vanderbilt tender at Bakersfield on September 23, 1954; it will be vacated here in November 1955. *(Harold Stewart: Menke collection)*

A visor-less headlight and distinctive ex-EP&SW class 70-SC-1 semi-cylindrical tender do nothing for the appearance of M-6 #1751. The oil bunker on the tender has been narrowed to allow better rearward visibility during switching moves. That tender once housed a coal bunker where the oil tank now resides. Photographed at Los Angeles on December 1, 1946, the #1751 would soon be transferred to the Tucson Division, where it served until June 1952. *(Allan S. Youell)*

At Santa Clara on September 23, 1950, Coast Division-assigned M-6 #1793 has only a few more months with a Pacific Lines identity before being leased to the SP of Mexico in March 1951. When the SPdeM was nationalized in December 1951, this engine would become Ferrocarril del Pacifico #422. This 7,300-gallon tender more nearly fits the definition of "semi-cylindrical" than the one behind #1751 in the image above. *(W. C. Whittaker)*

Once the war was over, railfans flocked back to excursions and SP gave them what they wanted. Still lettered in pre-1946 style at the Oakland Mole on May 4, 1947, M-6 #1795 waits to provide "rare mileage" over the backcountry San Ramon Branch. Perhaps for photographic authenticity, the sponsoring group (C-NRHS) requested an RPO instead of the usual baggage car. On the main line, this train ran as an extra, but train #60 will be inserted in the indicators once on branch rails—briefly reliving passenger service on this pastoral line. *(Arnold Menke collection)*

The pre-war "look" of small Southern Pacific steam locomotives is presented in 1939 at Los Angeles by M-6 #1798. It still carries a large headlight and the classic "Harriman" boiler-tube pilot, but—most importantly—has yet to be fitted with a power reverse. Instead, a long horizontal rod at the end of a tall lever in the cab reaches forward between the first and second drivers to reverse the Stephenson valve gear inside the frame. Engineers earned their pay on machines such as this, especially on switching jobs with frequent reverse moves.
(W. C. Whittaker)

While superficially similar to the M-6 class, M-9s #1804-1818 profited from engineering improvements and upgradings over the years. Alco built them in 1908 as part of a large Harriman Associated Lines order; in fact, #1804-1808 were delivered new to the Texas Lines (retaining their Pacific Lines numbers) and finally transferred west in 1910. Many served long and well south of the border, although none were sold to the SPdeM. In 1927, the #1807 received a large Worthington feedwater heater on the fireman's side at the SPdeM Empalme shop, forcing relocation of the air pumps to the right side.

In 1944, overflow freight from the Southern Pacific at Colton has come up to San Bernardino, where it enters the PE main line at F Street and Rialto Avenue. In deference to increased tonnage (and the PE's sawtooth grade profile), SP has provided Mogul #1811 as the road power, assisting home-built electric #1629. Joint SP-PE facilities in Pomona included a water supply for steam engines on this line.
(Norman Newcomb: Rasmussen collection)

A Western Division engine in late steam days, M-9 #1810 works at water side at beloved old Port Costa on Suisun Bay. The date is March 4, 1950, and #1810 has about three more years of active duty. *(A. C. Phelps: Menke collection)*

No, this huge train did not navigate Beaumont Hill with only M-9 #1808 as power; helping at the rear of the train is a more appropriate cab-ahead and another 2-6-0. It's November 16, 1952; passengers in the Pacific Electric bus bound for Redlands are going to have a long wait while this train rolls past. *(Stan Kistler)*

A stubby 7000-gallon Vanderbilt tender with tiny lettering will limit the operating range of M-6 #1823. This is one of the "war emergency" locomotives SP assembled in its own shops (from spare parts) during World War I, when it could not buy new machines from commercial locomotive builders. It served through that war and the next, but would finally be vacated in February 1950. In perfect light at Los Angeles on October 6, 1947, the #1823 has its oil bunker topped off. *(R. J. Berry)*

The six class M-11 locomotives, #1831-1836, arrived on the Pacific Lines in 1924 when the Arizona Eastern was merged into the SP; they were built by Baldwin in 1909. They weighed ninety tons and most were rebuilt with large cylinders and feedwater heaters. True to its heritage, the #1833 models the distinctive front ladders favored by the AE and SPdeM and still carries its old-style headlight at Los Angeles on October 9, 1946. It will serve another decade before being vacated from the roster.
(Leo Munson collection)

In 1939 and 1941, the SP leased ten 107-ton class M-21 moguls from the SP of Mexico, renumbering nine of them as #1837-1845. They were originally T&NO #520-529, products of Houston General Shops in 1928-30, and were considered to be the heaviest 2-6-0s ever built. Despite their impressive specifications, they were not a success on the SP; even with a crushing shortage of locomotives on the eve of World War II, SP returned them to the T&NO. The former SPdeM and T&NO #521 has lost its feedwater heater, leaving a large gap under the running board.

A clear steam exhaust from Texas interloper #1841 washes the de-electrified overhead on the main line at Fruitvale on February 19, 1941. Only a month before, Interurban Electric suburban passenger train service on these rails had come to a close. Westbound local train #437 is just a few miles from its terminal at Oakland Yard. Unlike the #1838 above, #1841 still carries its feedwater heater and pulls a Vanderbilt tender. In October 1942, the #1841 will be returned to the T&NO. *(Arnold Menke collection)*

Those AE-style front ladders appear again in a right-side image of #1835 (the former Arizona Eastern #574) at Los Angeles on January 16, 1948. The air pumps have migrated to the right side, crowding the power reverse, displaced by installation of a large Worthington feedwater heater on the left side. No Pacific Lines 2-6-0s were fitted with Walschaerts valve gear, although the T&NO class M-10 engines built in 1912 were treated to this modern feature.

Despite assertions that the modern—but overweight—class M-21 2-6-0s erected by T&NO's Houston Shops were built with all-new parts, including cast engine beds, the official erecting drawing refutes this notion. Instead, the T&NO apparently salvaged the chassis of ten class M-7/M-8 locomotives (built with Vanderbilt boilers, a discredited technology) that were all scrapped at Houston in 1928-29. When feedwater heaters were added, their weight on three driving axles jumped to 185,000 pounds—as compared to 147,000 on an M-9. They were tried and found wanting on the SPdeM's light rail and then moved to the Pacific Lines, which also failed to find a comfortable slot for them to fill. In August 1941, before returning to the T&NO and resuming its original number (#520), the #1842 rests on the Taylor Roundhouse ready track in Los Angeles. *(G. M. Best)*

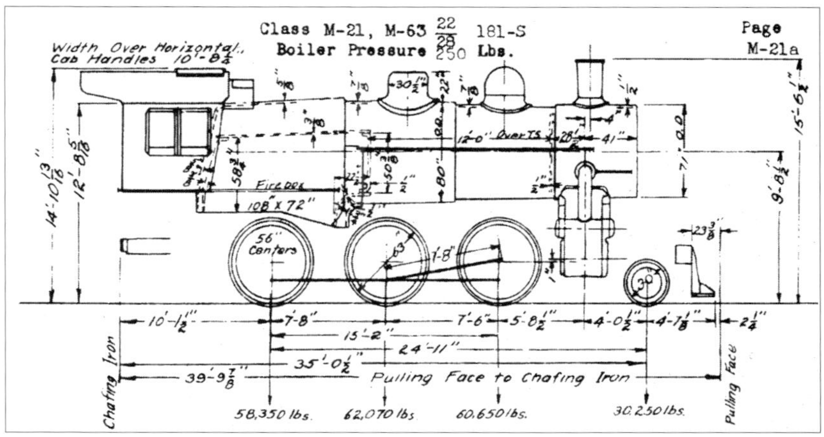

Track profiles on the relatively flat Pajaro and Salinas Valleys allowed dispatching fair-sized trains behind small locomotives. Ten-wheeler #2301, a Baldwin 1903 product in class T-23, has the *Gilroy Local* well in hand at Eaton siding, between Watsonville Junction and Aromas, on the last day of 1945. *(Fred Stoes)*

Chapter Four
Ten-Wheelers: 4-6-0 Type

Southern Pacific kept a few 1880s-era ten-wheelers around to perform light switching and fire train service, but those four class T-8 and T-9 machines were vacated by the late 1940s. Six class T-1 machines, built in 1895-1897 by Cooke and Schenectady, remained on the roster in the 1940s: #2242, 2248, 2252, 2267, 2269 and 2271. Three former El Paso & Southwestern 4-6-0s occupied second numbers in the 2100 series; they were used on Rio Grande and Tucson Division mixed trains.

Nine 1901-02 vintage Baldwin class T-26 engines numbered between #2283 and 2299 (built as Vauclain compounds) remained on the roster until the late 1940s. The two other oddities were #2385-2386, small engines leased from the San Diego & Arizona Eastern.

The bulk of the ten-wheelers, #2301-2384, were very much of a family design, despite being built over the time span between 1903 and 1920. The ten class T-23 locomotives, #2301-2310, were built by Baldwin in 1903 and paid for with Union Pacific money but were delivered before the actual Common Standard machines that were to follow. Built with slide valves, they were converted to piston valves and superheat by the 1920s. Five of them were sold in 1951 to the Ferrocarril del Pacifico, while the other five were retired between 1949 and 1957.

The T-28s, #2311-2352, were delivered between 1907 and 1911 with orders split between Baldwin and Alco. They were of true "Harriman Common Standard" design, built with piston valves but not factory-equipped with superheated steam. The Oregon Railroad & Navigation Co. (UP) received identical locomotives. In the 1917-21 era, they were rebuilt with superheat. Typical of Common Standard engines, they were easily maintained "workhorses" that remained on the roster until administratively retired, although fourteen of the original forty-two were sold in 1951 to the Ferrocarril del Pacifico.

Once the SP and UP were separated in 1913, plans went through for SP's own design of ten-wheeler, featuring piston valves, superheat and Walschaerts valve gear—as well as 69-inch drivers and larger cylinders. Alco built eight T-32s in 1913, #2363-2370. As World War I locomotive shortages set in, SP managers realized that they could not purchase their own designs of locomotives from commercial builders—and "not invented here" was Holy Writ on the SP—so locomotive production in Company shops was again started up (after dormancy for decades). Fourteen T-32s, essentially identical to the 1913 Alco machines, were delivered by Sacramento and Los Angeles shops between 1917 and 1920 so that SP could avoid purchasing USRA standard locomotives.

In May 1927, T-32 #2371 was rebuilt with a feedwater heater and additional improvements and reclassified as T-40, the only one of its class. Dear to the hearts of SP railroaders, the T-32 class remained unbroken until 1953, at the sunset of steam operation. However, not a one was preserved.

Last of the "Big Four," Collis P. Huntington essentially owned the Southern Pacific, but he died on August 13, 1900. He chose not to deal with the Baldwin Locomotive Works, so during the 1880s and 1890s, the SP bought few Baldwin locomotives. Immediately upon Huntington's death, however, Superintendent of Motive Power Henry Small turned to Baldwin for new locomotive orders, including 18 Vauclain compound 4-6-0s, #2283-2300. In only five years, however, this technology was proved to cost more in maintenance than it saved in fuel and the T-26 class was converted to simple cylinders. The #2283 is pictured in 1940 at Los Angeles, some years before it was leased to the Northwestern Pacific. *(G. M. Best)*

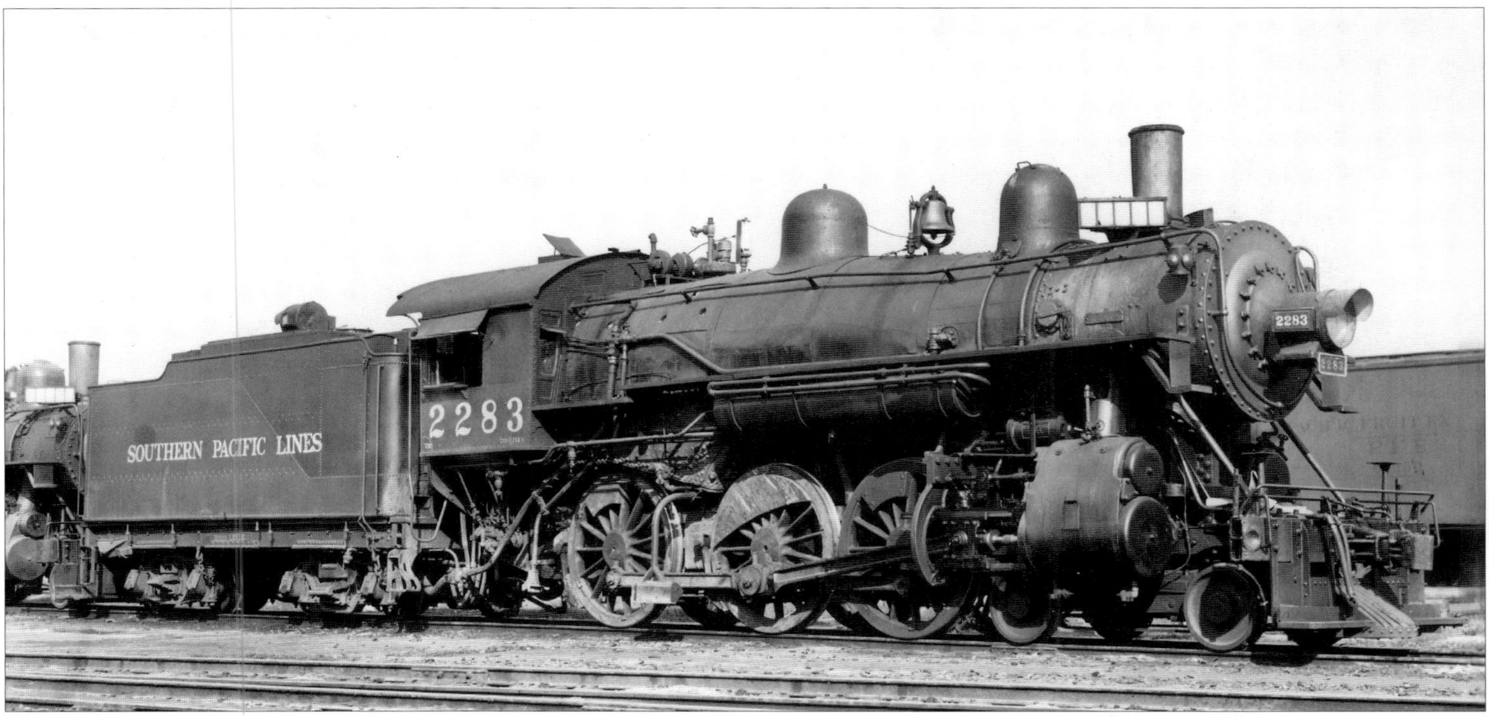

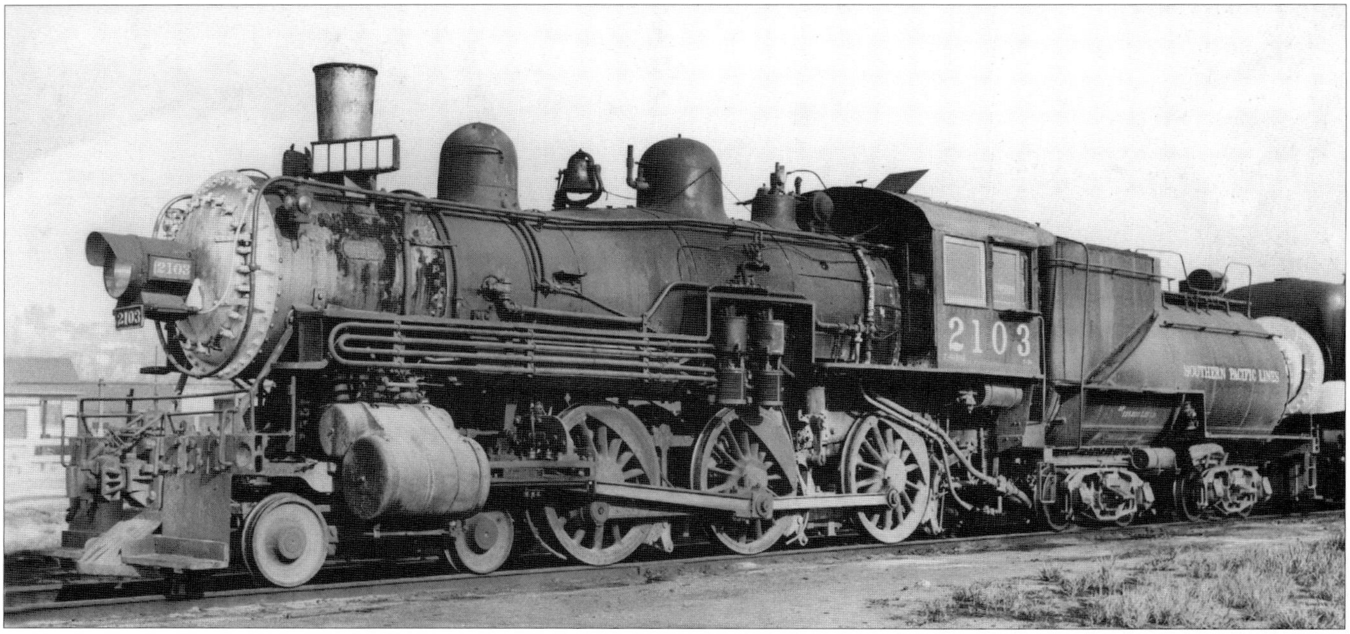

The only products of Alco's Manchester works on the entire Southern Pacific were the three inside-valve class T-36 ten-wheelers, #2102-2104, built in 1902 and inherited in 1924 from the El Paso & Southwestern. The #2102 and 2104 left the roster in the mid-thirties, while the #2103 (previously EP&SW #115) served its later years on the San Diego & Arizona Eastern. This engine was vacated at San Diego on January 13, 1947 and towed to Los Angles, where Gerald Best snapped a farewell image that February 11. In April, SP employees would cut it up for scrap.

The #2105 and 2106 were built for the El Paso & Southwestern in 1903 by Alco-Rhode Island. Originally #125-126, their owner rebuilt them several times in the next two decades—they ended up with 70-inch drivers and 24-by-28-inch cylinders. When they were absorbed into the SP roster in November 1924 as class T-37, they were found to be less capable than SP's own T-32 class and were banished to power Arizona passenger locals, where the #2105 was pictured in 1937 at Tucson. Both were supplanted by GE 70-ton units and scrapped in 1952. *(Allan S. Youell)*

The 4-6-0 locomotives that became SP #2114-2208 were built in large batches between 1885 and 1888. Schenectady, Cooke, Rhode Island and the Sacramento Shops all built their share. All weighed about 50-60 tons and were not particularly powerful, so they were of little use once the much larger Common Standard engines began arriving. Classes T-9 (#2154-2173) and T-8 (#2174-2183), both Schenectady 1888, were decimated during the Depression years when SP scrapped hundreds of locomotives. Only three survivors worked through World War II: #2161, 2174 and 2178. The final assignment for #2161 was yard work at Crescent Lake, where it was photographed in October 1948. *(Guy L. Dunscomb)*

One of the few named SP accommodation trains, the *Tomahawk,* operated between Globe and Bowie in Arizona. Departing Globe in the late morning of February 22, 1946, class T-37 #2105 has a small train (a postal-baggage car from class 60-BP-15 and a non-air conditioned coach) for the four-hour, 124-mile trip to connect with the main line at Bowie. For a number of years, trains #381-382 were routinely powered by former EP&SW ten-wheelers #2105 and 2106, 90-tonners built in 1903 by Alco's Rhode Island Works.

(Phillips C. Kauke)

At Sparks on August 3, 1946, T-8 #2174 is coupled to an auxiliary water car, suggesting use on a branch that had limited water facilities. Only two of this class, #2174 and 2178, remained on the roster in 1940—and then only because they were leased out. Never really modernized, #2174 features a switcher pilot, pedestal headlight and small rectangular tender. After a brief sojourn in Nevada, #2174 would be leased for a few months to the Northwestern Pacific, but was stripped for parts at Brooklyn, then scrapped there in August 1948.

(Douglas S. Richter)

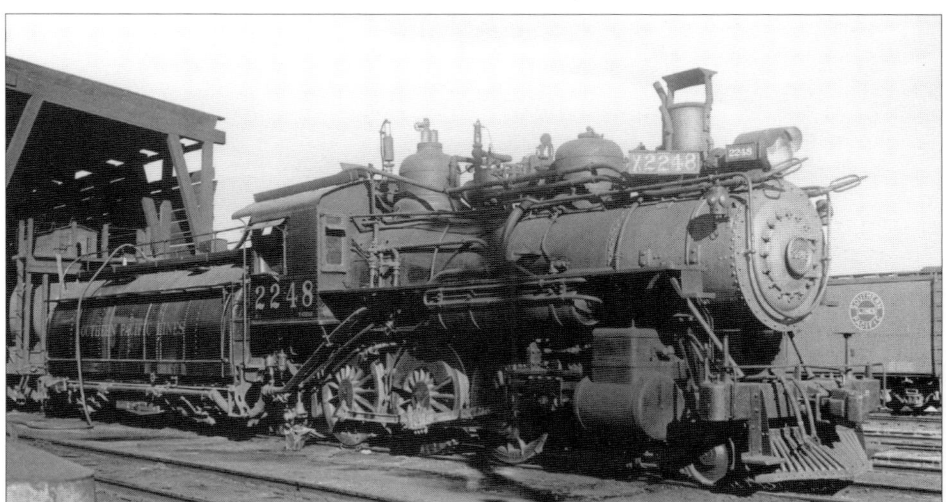

Southern Pacific's idea of a "fire engine" was a steam locomotive kept hot during the dry season, coupled to two water cars. This short train was ready to go on a moment's notice if and when fire erupted in the miles of wooden snowsheds in the Sierra. Mounted atop T-1 #2248's boiler (at the Fire Train Shed at Truckee in 1939) are a reciprocating water pump and a variable-tone steam whistle, ancestor of today's sirens. Fixed nozzles to spray the inside of snowsheds are mounted on the smokebox front. T-1s #2242, 2248 and 2252 enjoyed a long service life due to their assignment to fire prevention.

(Guy L. Dunscomb)

Two of the three T-1s (#2267 and 2271) that remained on the roster into the late 1940s were leased in 1947 to the Northwestern Pacific. The third, #2269, was assigned light switching and local jobs on the Los Angeles Division. As photographed at Taylor Roundhouse in 1942, it was equipped with SP's standard light-dispersion shield on the headlight. The #2269 would be vacated and scrapped in the spring of 1947 at the nearby Los Angeles General Shops. *(G. M. Best)*

Eight T-26 engines, half the original class, remained in service through World War II. Included among these 1901-02 vintage Baldwins were the #2283, 2289, 2292 and 2294-2299, many of which owed their longevity to the fact that they "fit well" on the Northwestern Pacific. For instance, the #2298, photographed on December 8, 1946 at Los Angeles, would be leased to the NWP during much of 1947-48 before being scrapped in 1949. *(Allan S. Youell)*

Longtime snowshed fire engine #2248 (Cooke 1896) was rebuilt in July 1955 to become SP's "old engine" for display at special events during the railroad's centennial celebration that year. Sacramento shops did a credible job (slightly overdone with an inappropriate stack and headlight), giving the Public Relations department what they wanted as well as a significant life extension to a small, old locomotive—one that remains operational on a tourist railroad in 2010. The #2248 is pictured at Sacramento during August 1955.
(D. S. Richter)

Perhaps its last picture on home rails, class T-23 #2309 pauses at Gila Bend on December 29, 1946. By this date, it has been equipped with both superheat and feedwater heater, as well as a power reverse installed in early 1940s. The left-side mounting of the feedwater heater displaced the air pump to the engineer's side. The ten class T-23 engines, #2301-2310, were built by Baldwin between August and October 1903. In two weeks, #2309 will report on lease to the SP de Mexico, remaining there until the railroad was nationalized in 1951—then becoming Ferrocarril del Pacifico #514.

Ganesha Junction was located northwest of downtown Pomona, more-or-less on the alignment of today's White Avenue in Pomona and not far from the PE's station for the Los Angeles County Fairgrounds. Part of the old SP line between Bassett and Pomona was used by the PE and finally sold to them. SP steam was thus running under PE overhead wire before World War II, demonstrated by ten-wheeler #2307 at Ganesha Junction on September 28, 1941. *(Phillips C. Kauke)*

Seven years after the image above, T-23 #2307 has migrated north from southern California, in the process acquiring a class 90-R-7 rectangular tender, switcher pilot and a Pyle headlight. It was photographed on January 25, 1948 at Watsonville Junction. *(R. C. Gray)*

Between 1907 and 1911 the Associated Lines purchased forty-two class T-28 locomotives for SP's Pacific Lines, numbered 2311-2352. They were built by both Alco-Brooks and Baldwin and—like most "Harriman" locomotives—were used until the end of steam operations. Thirteen (including Alco #2334, at Los Angeles on November 26, 1948) went with the SP de Mexico when that line was sold in late 1951 to the Mexican government. *(W. C. Whittaker)*

Final assignments for Baldwin T-28 #2315 (built in 1907) found it working in the Bay Area before being vacated in January 1953. A local freight trails #2315 at the Santa Clara Yard, near San Jose, in 1952. *(Arnold Menke collection)*

Ten-wheeler #2345 (class T-28; Baldwin 1909) leads three Common Standard coaches on street trackage in South San Francisco that never hosted a scheduled passenger train. The occasion was "A Day in San Francisco" fan trip on October 4, 1953, jointly sponsored by the Northern California Railroad Club and the Central Coast Railway Club.

Relatively low bridge load limits and light rail limited the size of locomotives that could be used on the San Diego & Arizona Eastern. Mikados and Pacifics were too heavy, forcing the SD&AE to do its job with 4-6-0 and 2-8-0 locomotives until SP could spare enough diesels to re-equip the line. Final dieselization was still three years in the future at San Diego on October 23, 1949, when exhaust deflector-equipped #2330 posed on the ready track. *(Stan Kistler)*

Stored on the "Lidgerwood Track" at Los Angeles on August 21, 1955, class T-31 #2355 has been put away with main rods disconnected in anticipation of next year's Imperial Valley vegetable rush season. Locomotives that were expected to see future use had their steam turrets bagged and (often) their cabs boarded-up. This engine was formally vacated almost a year after [illegible] [illegible]vice was closed out—and would be donated in 1958 to the city of Mesa, Arizona. *(Joseph A. [illegible]*

Originally planning to provide the San Diego & Arizona Eastern with diesels of its own, parent SP e[illegible] [illegible]ease roadswitchers to its subsidiary. However, full dieselization was still in the future in 1952, because SP had yet to p[illegible] [illegible]t diesels to protect helper assignments. At Dos Cabezas in 1952, year-old Baldwin-Lima-Hamilton AS-616 #5248 is the [illegible] [illegible]ith help from forty-year-old Baldwin 4-6-0 #2357. *(R. Neill: Menke [illegible]*

Heavyweight steel suburban cars arrived on the Commute service in the 1920s, forcing SP to assign high-driver 4-6-0s (and 2-6-2s from the EP&SW) to supplant less powerful eight-wheelers. In later years, the ten-wheelers held down only the less demanding off-peak schedules until they in turn were supplanted by 4-6-2s. San Francisco-bound Commute train #147, powered by #2358, is pictured in 1945 at College Park. This afternoon train carried a working RPO. *(Arnold Menke collection)*

Perhaps one of the most effectively illuminated locomotive photos to appear in this volume, an image of #2353 at the Bayshore roundhouse on June 15, 1954 benefits from diffused light as the afternoon sun sinks behind the Bay Area's perennial summer fog bank. *(Douglas S. Richter)*

SP subsidiary San Diego & Arizona Eastern operated Pullman and coach service from the Santa Fe station in downtown San Diego to Yuma via the Inter-California through Baja California, where connection was made with SP's *Golden State*, allowing the sleepers to go all the way to Chicago. T-32s #2383-2384 were leased to SD&AE in November 1939 as that road's principal passenger locomotives, but were returned to the SP after SD&AE passenger service was discontinued on January 11, 1951. Typically double-headed, eastbound train #362 is lined up at the San Diego Union Station on May 18, 1946. *(Stan Kistler)*

Springtime is coming to the Sierra and T-32 #2378 (at Roseville on April 17, 1949) won't need this interesting small plow pilot for at least the next few months. The exhaust stack deflector has been tipped back to allow a better draft. In the postwar years, SP dispensed with cylinder and valve cover "caps" as seen on the #2384 above—but already removed from the #2378. *(Allan S. Youell)*

An early Common Standard rectangular tender, blanks in the train number indicator boxes and a switcher pilot combine on May 31, 1949 to suggest that T-32 #2369 (Alco-Brooks, June 1913) is confined to local service out of Sparks, Nevada. In fact, the #2369 spent the rest of its career in the Reno-Sparks area, vacated here on September 2, 1954. It would be sold for scrap on November 2, 1954 to Purdy Metals in Los Angeles. *(Allan S. Youell)*

During World War I, power-short SP had endless problems obtaining new locomotives from the Eastern builders, forcing them to return to building their own. Using the same plans provided to Alco in 1913 for T-32s #2363-2370, Sacramento and Los Angeles Shops erected #2371-2384 between 1917and 1920. In 1927, the #2371 was fitted with a 3-BL feedwater heater, adding twelve tons to its weight, making it slightly heavier than the Pennsylvania G5s ten-wheelers, but not as powerful. Assigned in later years to the San Joaquin Division, T-40 #2371 was photographed on February 2, 1952 at Bakersfield. *(Allan S. Youell)*

The first section of San Joaquin Division freight #782 blasts through Selma, well on its way from Fresno to Bakersfield, on November 30, 1948. In this image, class T-40 #2371 has already received a switcher pilot, but the cylinder and valve covers have not yet been removed.
(Norman W. Holmes)

There was only one class T-57 on the Southern Pacific roster and it was considerably smaller than any of SP's "native" ten-wheelers, weighing just 81 tons. The #2385 was the former San Diego & Arizona Eastern #20, a hybrid boasting a Baldwin chassis and an SP-built boiler, operating with this SP number only from June 1941 until September 1948. Back at home on May 18, 1946 at the SD&AE engine house, #2385 would regain its old number in two more years and be scrapped at Los Angeles in September 1950.
(Stan Kistler)

This locomotive, built to serve the Nevada silver rush as Bullfrog-Goldfield #14, blew up in December 1910. The chassis was sold to the SP during World War I when SP was desperate for locomotives; it was rebuilt with a new boiler in December 1918 at Los Angeles General Shops and assigned to San Diego & Arizona #20. SP leased it back in June 1941, assigning it #2385. Back on home rails but not yet renumbered back to SD&AE #20, it is working a huge train at San Diego on July 29, 1947. *(Stan Kistler: Menke collection)*

One of many candidates for "shortest service in SP lettering" was the former San Diego & Arizona Eastern #26, a 94-ton ten-wheeler built by Baldwin in 1907 for the Las Vegas & Tonopah. SP leased it from the SD&A in April 1928 and finally took ownership in June 1941, assigning it #2386 and class (of one) T-58. In September 1948, after World War II, SP sold it back to its subsidiary. This photo was apparently taken in the San Joaquin Valley.

On June 28, 1955, fire engine #2248 operated on a test run and was photographed between Auburn and Emigrant Gap. This was apparently its final bow in its longtime service (with near-twin #2252) as an emergency resource to fight snowshed and tunnel fires. The Southern Pacific's oldest operable steam engine, T-1 #2248 already had a shop date to be "backdated" in appearance for the railroad's forthcoming centennial celebrations later in 1955. *(Don Ball collection: SP Historical & Technical Society)*

"I wish I'd done that..." does not apply to the young Gordon Glattenberg, leaning out of the cab of fast-accelerating P-10 #2484 in the very closing days of SP's mainline steam passenger operations—on September 12, 1956 at Lathrop. Train #54, the beloved Sacramento Daylight, has but two heavyweight cars today—no challenge at all for a powerful 4-6-2 and an enthusiastic engine crew.

(Bob Trennert: Glattenberg collection)

Chapter Five
4-6-2 and 4-4-2

By the 1940s, Pacific-type locomotives were second-tier passenger power, but they carried all the benefits of Common Standard engineering. Class P-1 (#2400-2427) arrived in 1904-07 featuring Stephenson valve gear, saturated steam (but with piston valves) and inside-bearing rear axles. Class P-3, delivered in 1911, adhered to basically similar specifications; these engines were numbered 2428-2437. The P-1 and P-3 classes were rebuilt with superheat and increased boiler pressure between 1917 and 1922.

By the time the class P-5 arrived (#2438-2452, Baldwin 1912), SP was ready to embrace factory-installed superheating as well as large valves and Walschaerts valve gear. While the basic specifications mirrored the P-1, SP obtained a much more capable locomotive with the P-5. In 1926-27, all of them had their boiler pressure re-set upward, instantly providing more tractive effort.

In 1927-29, Southern Pacific upgraded ten older 4-6-2s at Los Angeles and Sacramento with new cylinders, valve gear, Delta trailing trucks, boosters, feedwater heaters and sport cabs. They were reclassified to P-4.

The ultimate Common Standard 4-6-2 was the class P-6, distributed throughout the Associated Lines (even to the Chicago & Alton) and occupying #2453-2458 on the Pacific Lines roster. They were built by Alco-Brooks in May 1913. Wheelbase dimensions replicated earlier 4-6-2s, but both the firebox and the new wagon-top boiler were much larger, supporting 3-inch larger cylinders. The #2455-2457 went to the T&NO in 1937, where they were extensively modernized and streamlined for service on the *Sunbeam* and *Hustler* trains. Stay-at-homes #2453, 2454 and 2458 became fixtures on the Coast Division; in their latter years they carried air horns and pulled short tenders.

After World War I, SP designers stepped back four inches from 77-inch drivers and added two more inches to piston stroke, realizing the ten-ton-heavier P-8 (#2461-2475, Baldwin 1921) that could power heavy Overland Route trains. Quickly upgraded with feedwater heaters, the P-8s had their boiler pressure upped to 210 pounds in the 1939-42 era. They were vacated in the years 1953-57, but both the #2467 and 2472 have been restored.

The peak of "native" SP 4-6-2 development occurred in 1923-24, even as the 4-8-2s were on order. Baldwin built class P-10s #2478-2483 in October 1923 with feedwater heaters, then erected #2484-2491 in February 1924 with feedwater heaters and boosters. In 1941, the #2484-2486 were streamlined and painted for service on the new *San Joaquin Daylight*. Skyline casings were installed on #2481-2483, 2487-2489 and 2491 in the 1939-48 era. All saw heavy use during the war years, but were eventually demoted to Commute service. *(Continued on page 68.)*

Between 1904 and 1907, the Harriman Associated Lines allocated 28 class P-1 4-6-2s to the SP, #2400-2427, from both Alco and Baldwin. Virtually identical machines were assigned to various UP properties, the T&NO and even the Chicago & Alton. They were obsolete within a few years after they arrived, featuring Stephenson valve gear and the Baldwin (radial) Rushton "trailing truck" with very limited swing, not over 2" each way. Superheating was added between 1917 and 1920. Featuring a slightly off-center headlight, Baldwin-built #2412 is pictured at Bayshore in 1949 and would be scrapped in 1951. *(Guy L. Dunscomb: Richter collection)*

Trains #327-328, the *Shasta*, operated between Grants Pass, Oregon, and Dunsmuir, California, between 1946 and 1952, losing named status when the *Shasta Daylight* began operating via the Cascade Line. The Grants Pass turntable was too short to accommodate a P-1 with a Vanderbilt tender, so SP assigned #2407 (trailing a rectangular, 9000-gallon tender) and accepted an additional stop for water. On the very last day of operation, February 27, 1952, train #327 stops at the Medford station. *(John C. Illman)*

Three suburban coaches trail P-4 #2424 on San Francisco-bound Commute #149 at Santa Clara in 1949. The vertical pipe ahead of the stack is the trailing-truck booster exhaust. Well-qualified (after rebuilding to class P-4) for demanding Commute duties, #2424 features a corrugated sheet-steel pilot, auxiliary air reservoir and a feedwater heater. *(Arnold Menke collection)*

Rather than leave the P-1 4-6-2s in as-built condition, SP took steps to improve their performance. The entire class was rebuilt with superheat between 1917 and 1920, while #2400, 2406 and 2407 were modified in the 1920s with Delta trailing trucks and boosters, as well as a left-mounted #3-B feedwater heater—but they still retained their original Stephenson inside valve gear. The #2400 would be scrapped in October 1948, but was still ready for useful work on December 31, 1947 at the West Oakland roundhouse.

Still more performance could be gained from the basic P-1 boiler. In 1927, Sacramento Shops rebuilt two prototype P-4s, #2422 and 2409 (as well as two A-6s, #3000-3001) before turning over the project to Los Angeles Shops to complete. Both new classes utilized the original Common Standard boiler, but added a complete new cylinder and valve saddle, sport cab and Walschaerts valve gear atop the previous Delta trailing truck, booster and superheat. P-4 #2410 emerged from the Los Angeles Shops in February 1928 a much more capable machine. It is pictured on May 22, 1952 at the Mission Bay roundhouse in San Francisco. *(W. C. Whittaker)*

Morning sun angles on the Peninsula were unfavorable to train photographers, so it seems that many surviving images of the Commute trains were exposed in the afternoons. Train #151 was scheduled to depart San Jose at 6:15 in the evening, in bright sunlight during Daylight Savings Time. In the summer of 1941, P-3 #2429 escorts the two baggage cars and two coaches comprising today's train #151 past the herder's shed just west of SP's Cahill Street station in San Jose. *(Arnold Menke collection)*

Arriving in the midst of SP's steel car revolution, Espee's ten original P-3 locomotives, #2428-2437, were built by Baldwin in 1911. They followed the archaic P-1 blueprint, even to the extent of using saturated steam; about their only innovation was the adoption of the outside-bearing Hodges trailing truck. All were superheated between 1917 and 1920, while the #2436 was rebuilt in 1928 to P-4 specifications. Obsoleted by debut of the P-6 in 1913, they hung on until the late 1940s. The #2432 is pictured at the West Oakland roundhouse.

Hobbled by a small boiler and 19th-Century valve gear, the peas-in-a-pod P-1 and P-3 class locomotives were quickly bypassed by technology, but they remained on the roster until shouldered aside by Baldwin diesel roadswitchers. Coupled to a rectangular 9000-gallon tender, #2433 waits on the outbound ready track at Taylor Roundhouse in Los Angeles on the afternoon of November 26, 1947. It will be scrapped in 1952. *(G. M. Best)*

By the early 1950s, the T&NO owned more serviceable steam passenger locomotives than were needed for its remaining local passenger trains. Consequently, a lease was arranged in June and July of 1953 that moved four T&NO P-5 locomotives to the Pacific Lines—retaining their T&NO numbers. Most of their use was on the Western and San Joaquin Divisions. Pacific Lines P-5s #2438-2452 were contemporary, but the T&NO engines (#600, 602, 605 and 606) had been rebuilt in 1925 with Delta trailing trucks and boosters in the same manner as the Pacific Lines' P-4 rebuilds. Carefully stored with weather protection bagging the steam turret ahead of the cab, #602 waits at Tracy for the next busy season.

Time is running out for the peacetime Southern Pacific as San Francisco-bound train #179 pulls out of the new Palo Alto station on September 1, 1941. Two 96-seat Interurban Coaches and a mainline coach make this a light train. A light 4-6-2 was adequate power for an off-peak Commute such as this, but a heavy 4-6-2 with a booster would be required for an eight-car rush hour train.

Clearly reluctant to try anything new in 4-6-2s, the Associated Lines' designers took few risks when evolving the new P-5 class in 1912, the first year that all SP locomotive orders were specified with Walschaerts valve gear. The same basic P-1 chassis and boiler was specified once again, at least this time with factory-installed superheat, and finally adopting outside valve gear. Fifteen P-5 locomotives, #2438-2452, were built by Baldwin in August 1912. The #2440 stops off at the Modesto freight station around 1949, still wearing its cylinder and valve end covers.

At West Oakland in 1939, P-5 #2450 exemplifies the "old look" of Southern Pacific 4-6-2s: no rain gutter strip on the edge of the cab roof, a pedestal headlight, 9000-gallon rectangular Common Standard tender and small "Lines" lettering on the tender.

(W. C. Whittaker)

Once the war emergency was over, shop forces had the relative leisure to modernize features on older locomotives. The post-1946 "new look" included a rain gutter strip on the edge of the cab roof, a small, visored Pyle-National headlight, large SOUTHERN PACIFIC tender lettering and an aluminum-painted smokebox door to enhance visibility. The downside was that locomotives were less frequently washed, displaying ample proof that not all boiler water was of top quality.

The Harriman regime's final design for a 4-6-2 was a huge step forward in terms of locomotive design. Finally adopting a two-foot-longer chassis and a much larger wagon-top boiler (boasting 144% more heating surface than the P-5), the P-6 could easily generate sufficient steam to feed three-inch larger cylinders. Built by Alco in spring 1913, P-6s were distributed to the SP (#2453-2458), T&NO (#610-621), the UP and even the Chicago & Alton. As the SP's P-6 engines were entering service, the courts were severing UP control of the SP. With a clear stack, #2458 breezes into Burlingame on April 18, 1952. *(A. W. Styffe)*

Tracy-to-Oakland train #401 must surmount Altamont Pass and Niles Canyon before reaching the Bay Area, so officials at Tracy have opted for a run-through helper to operate at least as far as Radum. In March 1955, P-8 #2465 is doubleheaded with 4-8-2 #4341 as #401 departs the Tracy yard. Both engines had seen better days as passenger power, but SP was by now simply running off their mileage before Electro-Motive could deliver new GP9s. *(John Harder: Menke collection)*

A consist enlarged by the crush of holiday travelers finds the six-car *Del Monte* outbound from San Francisco at Newcomb Avenue on the afternoon of January 1, 1952. In charge today is #2454, an Alco-Brooks product from 1913, one of the three P-6 regulars on this train. *(John C. Illman)*

Perfect late-afternoon light illuminates P-6 #2453 as it stops for water at Watsonville Junction on April 27, 1955. The train is the iconic San Francisco-Monterey *Del Monte*, SP's last passenger service dedicated to the carriage trade. Short tenders were required to fit on the turntable at the end of the line, but literally as this photo was snapped, new GP9 passenger diesels were on their way from Electro-Motive and the #2453 will be vacated in September. *(Ray Whitaker)*

During World War I, commercial locomotive builders did not have the capacity to serve the war effort and still turn out "customized" locomotives for rail lines fueled by oil, essentially telling SP to "order after the War." SP needed more locomotives, so shops were mobilized to gather spare parts to build new examples in-house of 0-6-0, 2-6-0, 4-6-0 and 2-8-0 locomotives. Even a pair of obsolete-design P-3s emerged in 1917-18 as #2459-2460; they were capable within their design limitations and would serve until 1952 and 1948. Twenty-one-year-old #2459 exemplifies pre-World War II standards at West Oakland in November 1938.

(Arnold Menke collection)

Finally, twenty more 4-6-2s were added in 1924, purchased with the El Paso & Southwestern. The P-11s (#3100-3109, Baldwin 1907) were comparable to SP's own P-4 machines; they were converted from coal fuel in the 1930s and moved to California for Commute service, but they were not particularly successful. The #3103, 3105 and 3107 were all scrapped before the war, while the rest were all off the roster by 1951.

Class P-12, #3120-3129, was built in 1917 by Alco-Brooks. They were heavier and somewhat more powerful than SP P-10s; all were converted to oil fuel between 1925 and 1930. Rough-riding and "over-cylindered" as received, they were extensively modified and ultimately given smaller cylinders. Unpopular with engine crews, they were vacated between 1947 and 1953.

Baggage cars were not normally seen on Commute trains, but this one was being returned to San Francisco after shop work. Bell swinging, P-8 #2466 delivers westbound Sunday-and-Holiday-only Commute #179 into Santa Clara at 2:35 in the afternoon.

(John Harder: Menke collection)

Far past its prime as a first-line passenger locomotive, P-3 #2459 pilots the second section of San Joaquin Division freight #780 into the Bakersfield yard on the afternoon of February 13, 1951. Small, second-line locomotives could easily move freight across the San Joaquin Valley, but it will require a Cab-Ahead to forward this train over the Tehachapis. Thirteen months after this picture, shop forces at Los Angeles will close the book on this 1917 Sacramento Shops product. *(Phillips C. Kauke)*

Arrival of class P-8, #2461-2475, marked the beginning of a renaissance in SP steam locomotive design after years of Harriman stagnation and then United States Railroad Administration control during World War I. Built by Baldwin in 1921, these robust machines addressed boiler capacity (power) issues by trading boiler length for a combustion chamber. A longer cylinder stroke and smaller, 73-inch drivers provided a 5,000 pound increase in tractive effort. As the first 4-6-2s delivered with the Delta trailing truck, the P-8s could accommodate a booster, but only #2475 actually received this feature. Allan Youell pictured the #2467 at Sacramento on May 31, 1948.

In the 1920s, when labor was inexpensive, SP could afford to keep its passenger locomotives glittering. Bare metal was wiped down and polished, trim was highlighted with contrasting paint or graphite and boilers were often painted in colors other than black. The economic woes of the Great Depression changed all that, but a few SP engines in high-visibility assignments remained in the grand old style, demonstrated at the Sacramento station on September 16, 1939 by P-8 #2474, marked up for Oakland-bound train #229.
(Herman Friedrich)

Shadows grow long at Shellmound Tower in Oakland (where the former Interurban Electric lines to Berkeley curved off to the northeast) on the late afternoon of June 30, 1941 as the eastbound *El Dorado* accelerates in the wake of P-8 #2472. Nine heavyweight cars, including a diner, underscore SP's need for powerful passenger locomotives to meet schedule requirements on the demanding San Francisco-Oakland-Sacramento service. The #2472 remains available for service in the 21st Century. *(W. C. Whittaker: Menke collection)*

At 6:54 every weekday evening in the early 1950s, regular Commute train #132 arrived at suburban Los Gatos, southwest of San Jose in the foothills of California's Coast Range. After running around the train, class P-7 Pacific #2477 will back up to San Jose as "equipment" train 193 with no passengers allowed. *(Charles Givens)*

In August 1917, Lima built two class P-7 4-6-2s for SP subsidiary Arizona Eastern as their #701-702. They followed P-6 blueprints except for 73-inch drivers substituted for SP's standard 77-inch, in deference to the AE's slower train speeds and heavier trains. They were sold to the SP in February 1922, becoming SP #2476-2477; both ended their days in Commute service and were still on duty at the close of steam operations. The #2477 waits at Third & Townsend in San Francisco on April 14, 1953, ready to go with train 130, due out at 5:19 p.m. *(Guy L. Dunscomb)*

Essentially the only difference between the Alco-Brooks-built P-6 locomotives of 1913 and the 1917-vintage Lima P-7s was their driver diameter; Arizona Eastern #701-702 (SP #2476-2477 after 1922) were built with 73-inch drivers (rather than 77) to haul heavy trains over Arizona's sawtooth track profile, where speed was relatively unimportant. Once they moved to the SP, their two thousand extra pounds of tractive effort was still useful. Already modernized when photographed on May 19, 1946 at the Brooklyn roundhouse near Portland, the #2476 displays its thick driver tires. *(Don H. Roberts)*

In 1941, three 1924-vintage P-10s, #2484-2486, were rebuilt with skyline casings, decorative "skirts" and repainted in red-and-orange. They were assigned to the new streamlined *San Joaquin Daylight* service between Oakland Mole and Bakersfield, inaugurated on July 4, 1941. In the early years of this train, SP routinely swapped out passenger power on the *Daylights* at Fresno; that's what is happening in this image at Fresno on June 14, 1941, as the depot switcher pulls #2484 to the roundhouse for turnaround servicing. The skirts were painted black during World War II and removed April-June 1950 at West Oakland. *(Allan S. Youell)*

Locked in hot competition with the Santa Fe for Valley patrons, the SP took great pride in its steam-powered streamlined *San Joaquin Daylight,* pictured at the Oakland Mole on the morning of September 5, 1941. For the 309 miles between Oakland and Bakersfield, P-10 #2485 (or one of its two streamlined siblings) was more than capable of maintaining the schedule—a job that would require a 4-8-4 on the hilly Coast Line. From Bakersfield south to Los Angeles, heavier power replaced the 4-6-2s. The Santa Fe had no direct rail route south from Bakersfield; its customers had to ride a bus to Los Angeles. *(Phillips C. Kauke)*

Baldwin delivered P-10s #2478-2491 in 1923 and 1924—the last of this wheel arrangement built for the Pacific Lines. SP needed more power for long through trains, turning next to Alco to provide the first 4-8-2s. The P-8 and P-10 locomotives held mainline passenger assignments (as well as Commute duties) until the end of steam operation, demonstrated by #2490 on train #54—the *Sacramento Daylight*—at Lathrop on May 1, 1954. *(Guy L. Dunscomb)*

Right at home among 4-8-4s at Fresno on May 13, 1955, P-10 #2490 will draw freight duties today. In this class, #2478-2480 and 2490 never received skyline casings, but the latter-day corrugated steel pilot is apparent in this view. *(John E. Shaw, Jr.)*

Sheet-steel skirting was removed from #2484-2486 in 1950, but they retained their distinctive front ladders and skyline casings. An LFM "Universal" disc main driver was installed on #2484 as an experiment in 1949, the only instance on the Pacific Lines. Reprising the two-car *Sacramento Daylight* consist featured on the opening page of this chapter, #2484 has in tow express-baggage (star car) #6042, class 60-B-9, and *Daylight*-painted combine #3176. *(Marvin Maynard: Menke collection)*

Markers have been hung on venerable Common Standard coach #1856 at San Jose, circa 1953, making this a "train," not a switching move—but otherwise the circumstances for a one-car Extra 2486 east are unexplained. Formerly streamlined P-10 #2486 displays a lower level of cosmetic attention than was typically applied to Commute engines, suggesting that, at this point in time, it was primarily used on freight trains. *(Arnold Menke collection)*

Baldwin used a patented "Rushton" trailing truck on its early 4-4-2, 2-6-2 and 4-6-2 locomotives. In its original form (as on the SP class P-1 engines), this truck was designed with an inside bearing that allowed only minimal "swing" on curves. After a few years, the truck was redesigned with outside bearings—as applied to El Paso & Southwestern 4-6-2s #140-149 in 1907. These machines became Southern Pacific class P-11, #3100-3109, in late 1924. Pictured on November 10, 1946 at Los Angeles, #3102 has both a switcher tender and switcher footboards and has less than two years of active service remaining. *(Allan S. Youell)*

Far from its original home, former El Paso & Southwestern #148—now SP P-11 #3108—is pictured on July 21, 1947 on the Geer Branch wye at Salem, Oregon, with the *Salem Local*. The ten 1907-vintage Baldwins in this class were rebuilt from coal to oil firing in 1929-30 and initially assigned to Commute duties in San Francisco. There they received rectangular tenders and other typically "SP" features. The #3103 and 3107 were vacated in 1939, but the remaining eight lasted through the War. *(W. C. Whittaker)*

With drivers measuring just 68 inches, the class P-11 locomotives were best qualified for local service and were used on the Los Angeles Division to deliver reefer blocks from remote locations to major icing stations—work often assigned to Mikados. Confined in their duties, the P-11s were rarely seen with large road-engine tenders. A modern low-profile pilot and a small "Common Standard" rectangular tender highlight #3106 at Los Angeles on the afternoon of June 6, 1947. *(Allan S. Youell)*

As was the case with the 3300-series Mikados, former owner El Paso & Southwestern ordered heavy Pacifics #160-169 from Alco with very large cylinders and relatively low boiler pressure—as well as coal fuel. SP inherited them in 1924 as #3120-3129 and quickly converted them to oil fuel. Feedwater heaters were added between 1927 and 1930, but still more rebuilding was required; cylinder bores were eventually reduced. Enginemen still did not like them. Vivid Oregon sunlight on July 2, 1949 at the Brooklyn roundhouse highlights Portland Division-assigned #3121 just three months before it would be dropped from the roster. *(Dick Wolf)*

A quadruple-header of two Pacifics, a Mogul and a Mikado (#3123, 1742, 2445 and 3236) struggle at twenty miles per hour with ninety-nine cars at Valley Boulevard in Alhambra on January 16, 1942. This occasion marked the low point of steam locomotive availability on the SP at the outset of World War II, when every imaginable combination of locomotives was put to work. *(G. M. Best: Menke collection)*

Once World War II was over, SP trains delivered thousands of military personnel across the country to separation centers and hospitals; Main Trains clogged the rails well into 1946-47. Showing signs of heavy war time use and (as yet) no postwar modifications, Los Angeles Division-assigned #3123 has been called for Beaumont Hill helper duties at Colton for extra 4358 east on February 9, 1946. The #3122, 3123 and 3127 were the last three of the ten class P-12 4-6-2s; they survived until 1952-53. *(Allan S. Youell)*

4-4-2 Locomotives

Four class A-6 4-4-2 locomotives, second #3000-3003, were rebuilt in 1927-28 from older class A-3 machines, utilizing every innovation SP designers could envision, in order to produce a modern engine for secondary service. After World War II, two received red-and-orange paint on their cabs and tenders and were assigned to the Sacramento connection of the *San Joaquin Daylight*. Leaving Lathrop for Stockton in 1949, booster-equipped #3000 has an eight-car train in tow. *(James E. Boynton)*

Before World War II broke out, class A-3 Atlantic #3052 reposed in semi-retirement at Calexico in stationary boiler service, but SP's desperate need for locomotives brought it back to work. Equipped with an uncommon 7300-gallon semi-cylindrical tender, this 1906 Baldwin (rolling on tall 81-inch drivers, but equipped with a booster) won't contribute much to the war effort, but power-short railroads couldn't afford to be picky. The #3052, photographed on May 9, 1942 at Taylor Roundhouse in Los Angeles, will serve until October 1949. *(G. M. Best)*

On Pearl Harbor Day, Sunday, December 7, 1941, it was still business as usual for the SP. Train 202, with a working RPO, is powered at Sacramento by booster-equipped A-3 #3036, already boasting an aluminum-painted smokebox. After the war, the #3036 would be rebuilt with a corrugated steel pilot, but finally scrapped in February 1949 at Sacramento. *(W. C. Whittaker)*

In the immediate post-Huntington years, SP experimented extensively with Baldwin Vauclain compounding. The first "new century" passenger engines, in fact, were Vauclain compounds in class A-1 (#3000-3015; Baldwin 1902) and A-2 (#3016-3024; Baldwin 1903), the latter group featuring Vanderbilt boilers. Neither class was a success; they were swept aside by new Common Standard 4-4-2 and 4-6-2 locomotives delivered beginning in 1904. The A-2s were SP's shortest-lived class, off the roster by 1915-18.

SP received Common Standard class A-3 4-4-2s #3025-3037 in August 1904 from Alco-Schenectady, then #3038-3065 in 1906 and 1907 from Baldwin. Similar machines were built for the Union Pacific and its subsidiaries, as well as the SP's Mexican and Texas operations. They were built with saturated steam, but carried piston valves actuated by Stephenson valve gear. They were reasonably powerful but were quickly obsoleted by the advent of steel passenger equipment.

The Harriman administration transferred six more A-3s to the Pacific Lines in 1912-13, where they became #3066-3071. In 1924, former Arizona Eastern #870-872 (Baldwin 1911) became SP class A-5 #3072-3074. In a program that exactly paralleled the rebuilding of class P-1 4-6-2s to class P-4, in 1927-28 SP heavily rebuilt four A-3s with large piston valves and Walschaerts valve gear, reclassifying them as A-6 and renumbering them second 3000-3003.

SP needed these light engines for Oakland-Sacramento runs because they could fit on the passenger train ferries that carried SP trains across Benicia Straits. As trains grew larger, however, during 1923-26 SP resorted to fitting thirteen with boosters to enhance their performance within existing weight limits. Once the new bridge was completed in 1930, SP was relieved of this weight bottleneck and most of the A-3s were suddenly out of work. Between 1932 and 1936, SP completely scrapped 24 of them, while dismantling eight more but salvaging their boilers to be installed on new 0-8-0 chassis.

Those A-3s that had been rebuilt with trailing truck boosters gained a performance edge allowing them to perform useful work for a few more years. Postwar survivors were few and far between: #3025, 3029, 3036, 3052 and 3057. The #3025 was donated to the City of Los Angeles and became one of the first locomotives exhibited at Travel Town in Griffith Park; it is still there.

The modernized A-6 locomotives served through the war years and after, with two (#3000-3001) being assigned to the streamlined *Sacramento Daylight* that delivered passengers from Sacramento to a connection at Lathrop with the *San Joaquin Daylight*. They were painted and striped in red and orange to match the lightweight cars. Even with this high-profile assignment, time caught up with them and they were retired in 1950.

The few remaining booster-equipped Atlantics were gathered on the Western Division, where they were kept available for off-peak passenger trains between Oakland Pier and Sacramento. However, as passenger trains grew larger and slower, there was less need for small, fast locomotives; furthermore, 4-6-2s were being "bumped" from mainline trains. The last two 4-4-2s would be vacated from the roster in 1952. The #3057 was photographed in 1947 at West Oakland.

Both Alco and Baldwin built Common Standard 4-4-2s numbered between #3025 and 3071, delivering them were between 1904 and 1908. Originally, the A-3s were fitted with Rushton radial trailing "trucks" similar to those on the P-1 class Pacifics. In this configuration, they could not start a heavy train, so many were modernized with Delta trailing trucks and boosters. Alco-built #3025, at Tucson on July 3, 1949, was rebuilt with a trailing truck booster in 1923. This engine would become the first SP steam locomotive to be donated to a municipal government for display, moving to Travel Town in Los Angeles' Griffith Park in 1952.

(John Lawson)

Backed in to Los Angeles Central Station on Alameda Street about 1916, #3064 illustrates a Common Standard (A-3) Atlantic as this class appeared in its early days. Virtually identical locomotives were doled out by the Associated Lines to the Union Pacific and its subsidiaries, as well as to the Texas & New Orleans and SP's Mexican subsidiaries. The #3064's Vanderbilt tender still rolls on arch-bar trucks, but the locomotive itself is already equipped with electric lights and train number indicators. It will be fitted with superheat in 1921; in the 1930s the lanky chassis will be scrapped, but the boiler will be salvaged at Sacramento shops to power new 0-8-0 #1314. *(Strapac collection)*

The Southern Pacific's "ultimate light passenger locomotives" were the four class A-6 machines rebuilt in the 1920s to better utilize their free-steaming boilers. Modernized with Walschaerts valve gear, large piston valves, Delta trailing trucks with boosters and sport cabs, they were an immediate success. Onset of the Depression stifled plans for any more. Officially vacated from the roster on February 28, 1950, A-6 #3001 actually steamed its last at the Mission Bay roundhouse on the afternoon of March 4; this picture was taken that very evening. The only reason for a Western Division-assigned 4-4-2 to appear in San Francisco (which had no operational requirement for such a small machine) was to provide a temporary source of depot steam. *Daylight* red-and-orange paint on the cab and tender is still evident. *(Douglas S. Richter)*

Extra 3234 west, near Puente, was photographed on Pearl Harbor Day, December 7, 1941. Not long after this, SP began installing "light dispersion shields" on headlights, signals and marker lights, as well as discouraging railfan photography. For the moment, however, it appears that 2-8-0 #3443, recently reclaimed from Mexico, is doing all the work–or perhaps the fireman is saluting the photographer by sanding out the flues. *(Phillips C. Kauke)*

Chapter Six
Consolidations: 2-8-0

The overwhelming "favorite steam locomotive" of the SP-T&NO was the class C-9 2-8-0. There were 169 C-9 locomotives on the Pacific Lines: #2513-2599, 2752-2830 and 2858-2860, built 1905-1908, a product of Mr. Harriman's "Common Standards" design team and rendered by the hundreds by both Alco and Baldwin for the SP, Union Pacific, Chicago & Alton and even the Erie. Harriman's 1904-1913 locomotive standardization program aimed at reducing first cost by placing large orders, easing maintenance via commonality of parts, and attaining the best possible performance within axle loading limits. Ultimately, the SP was to own over 250 Common Standard 2-8-0s in classes C-8, C-9 and C-10.

Historians concur that earlier SP 2-8-0s in classes C-2 through C-5 (Schenectady cross compounds and Baldwin Vauclain compounds) were the prototypes for Common Standard machines introduced in 1904, even though the Associated Lines wisely avoided use of compound steam. Until 1909, these 2-8-0s were the railroad's "big" road freight engines, leading trains everywhere between Portland and New Orleans.

Over the years, they were modified with superheat and larger tenders—and some fitted with prominent feedwater heaters—but none were rebuilt with modern valve gear. SP was to keep Common Standard 2-8-0s working until the very last days of steam freight operations, then fired up a few as stationary steam suppliers.

In 1924, the El Paso & Southwestern was leased to the SP. Included were 57 2-8-0 locomotives that became SP #3400-3426 and 3440-3469. Most of the smaller engines in the #3400-3426 group, classes C-18 and C-19, had been rebuilt with superheat and piston valves, retaining Stephenson valve gear. Ten did not survive until 1940.

Classes C-26 though C-29 (SP #3440-3469) were extensively rebuilt on the EP&SW, boasting large cylinders, valves and Walschaerts valve gear. With their modern features, they appeared to be more powerful than "native" Common Standard 2-8-0s, but in reality they were similar in performance and suffered the indignity of not being "invented here." They were converted from coal to oil firing on the SP of Mexico, where they operated until the Pacific Lines ran short of power during World War II. Many of them migrated to the Los Angeles and Tucson Divisions, where they were fitted with latter-day SP appliances. Pictures are relatively scarce, because by 1950 most (except #3447 and 3457, lost in wrecks) were back in Mexico—and were sold in 1951 as a group to the Ferrocarril del Pacifico.

After World War II, it was less expensive for SP to lease steam locomotives than buy new diesels, so in 1946, SP took three-year leases on seven Cotton Belt 2-8-0s for use on the Tucson Division: SSW #554. 560, 562, 563, 569, 574 and 575 (they were not renumbered or re-lettered, but they did receive aluminum smokebox fronts and train number indicators). Similar conditions applied ten years later when new diesels slated for the Coast Division were delayed; in early 1956, amazingly, SP again tapped the SSW for eight 2-8-0s—which this time *were* renumbered as SP #2861-2868.

The "do-everything, go everywhere" locomotive of the SP steam roster was the Harriman Common Standard 2-8-0, of which hundreds were delivered 1904-1907. Baldwin class C-9 engines in the 2500 series were allocated in more or less reverse delivery order, with #2587-2599 delivered in early 1906, while #2513-2535 didn't arrive until September-October 1907. The #2520, pictured at Los Angeles on October 23, 1949, had an odd, low stack to accommodate installation of an exhaust deflector. *(Allan S. Youell)*

Fitting right in with the former El Paso & Southwestern 2-8-0s illustrated below, 72-ton, slide valve *third* #2500 was a 1942 addition to the roster; SP bought it from the Nacozari Railroad in Mexico. The only representative of class C-32, it was built by Baldwin in 1906. In the brief six years that it served as SP #2500, almost no modifications were made, but it did fill a need during the War. It was used on the Northwestern Pacific for eleven months, but was scrapped early in 1948. *(G. M. Best)*

The original 2-8-0 locomotives occupying numbers 2500-2512 were off the roster by 1919, leaving a convenient "empty space" for SP to add small 2-8-0s acquired in later years. In 1924, four former El Paso & Southwestern 2-8-0s, #181-184, became SP second #2505-2508 (class C-15). They were built as coalburners in 1900, but in 1925-30 SP converted them to oil fuel, retaining their odd tenders but eventually re-assigning them to the Portland Division. Probably already vacated in this photo, the #2505 will be scrapped at Brooklyn in June 1949.
(Don H. Roberts)

Baldwin built two 88-ton 2-8-0s in 1902 and 1905 for the Arizona & New Mexico, an EP&SW predecessor; they became SP second #2510-2511 in 1924. Oil-fired class C-17 second #2511 is pictured in 1947 at Alamogordo. Although similar in appearance to the #2505 above, the #2511 weighed 15 tons more and boasted larger cylinders. A virtually "deckless" cab arrangement dictated the installation of a head brakeman's "doghouse" on the tender. The two C-17s were scrapped at El Paso within a few weeks of each other in November 1949 .
(G. M. Best)

The Big Trees area above Santa Cruz was always popular as a Bay Area rail excursion destination, both for railfans and picnickers. Trains originating in San Francisco used large passenger locomotives to quickly deliver passengers to Watsonville Junction, where locomotives more suitable for branchline service would replace the mainline power and the excursion trains would continue up the branch. At Rincon (near Felton on the former South Pacific Coast) on July 20, 1952, 2-8-0 #2598 and 2-8-2 #3227 toil uphill on the 1.45% grade with a train of Commute "subs." *(Norman W. Holmes)*

Reflecting the San Jose area's transition from farming to suburban housing and from production to consumption, SP freight deliveries dwindled and freight agencies were being closed. In the early fifties, extra 2558 east, the local freight, arrives at Los Gatos with only three cars and a caboose. *(Charles Givens)*

The #2513, the lowest-numbered (but not the oldest) C-9, was photographed on a cloudy day at Bayshore in September 1953. Only a few 2-8-0s were assigned large 12,000-gallon Vanderbilt tenders; this class 120-C-6 tender, when filled with water and fuel, weighed some twenty tons more than the locomotive.

The Harriman Associated Lines Common Standard CS-103 specification eventually covered some 690 2-8-0 locomotives allocated to the SP, UP, Erie, C&A and a few other roads. SP ended up with 289 of them in classes C-8 and C-9, with the #2599 going to work on the SP on March 2, 1906, the highest-numbered but not last delivered of the 2500 series. Assigned (with a short Vanderbilt tender) to the Portland Division in May 1948, it is parked at the Brooklyn roundhouse. In May 1950, it will move to Mexico. *(Don H. Roberts)*

Fully adapted to switching service with a "bobbed" Vanderbilt tender, no train number indicators and footboards in place of a road pilot, C-5 #2641 powers a local at Lawndale (almost directly east of Santa Rosa at Milepost 64.9) on October 7, 1941. Crew members were discouraged from riding the pilot beam in this manner, but perhaps the trainmaster was on vacation!
(W. C. Whittaker)

This 1900-vintage Schenectady 2-8-0, class C-3, was in line for retirement at the time this picture was taken at Portland on July 21, 1947; it would be vacated in February 1949. Very few SP steam engines (except old ones like #2610, assigned to the Portland Division) carried their headlights atop the smokebox into postwar years. *(W. C. Whittaker)*

Collis P. Huntington lived long enough to sign the order documents for modern locomotives, including six cross-compound Schenectady 2-8-0s delivered in April 1899 that became SP #2600-2605. These class C-2 machines were rebuilt to simple between 1910 and 1917 and then superheated between 1921 and 1929. The #2602, pictured on December 18, 1940 at Los Angeles, would be scrapped at Bayshore in March 1947 .
(G. M. Best)

The six class C-3 locomotives, #2606-2611, were built by Schenectady in 1900; #2608-2610 were superheated and survived into postwar years. While wheelbase dimensions were similar to the later Common Standard 2-8-0s, the C-2 and C-3 engines carried very long stroke (22x34-inch) cylinders and significantly smaller boilers than their successors. In October 1948, the #2609 was assigned to the small roundhouse at Weed, California.
(W. C. Whittaker)

Most of SP's 2600-series 2-8-0s were designed and built prior to Harriman Common Standard days, but their specifications anticipated the later machines. Class C-4, #2612-2623, came from pre-Alco Schenectady as two-cylinder cross compounds in 1901; they were rebuilt to simple between 1907 and 1916. Half of them received superheat and piston valves, but the #2617 soldiered on as a "soak" until it was vacated in May 1947. Despite fresh paint, the #2617 was only six months away from final retirement at Los Angeles in November 1946.
(Allan S. Youell)

Built by Baldwin in December 1901 as a Vauclain compound, class C-5 #2630 was rebuilt to simple operation in March 1906. However, the next logical step-superheated steam-was never installed and the #2630 was doomed to run out its years as a "soak" with slide valves. Kept on the Coast Division as a yard switcher, the #2630 has been fitted with switcher footboards and one of SP's unique semi-cylindrical tenders. It will be scrapped in 1949 at Sacramento, a victim of advancing age and deliveries of diesel switchers.

Class C-5 #2676 is pictured in April 1949 at Eugene, fitted with switcher footboards and trailing a Vanderbilt tender with a "cropped" oil tank for better rearward vision in switching moves. While it still carries train number indicators, most of its work by now is conducted within yard limits. Built in 1903 by Baldwin, this Portland Division-assigned switcher will be vacated in June 1949 and scrapped the following month at the Brooklyn shops.
(W. C. Whittaker)

On June 21, 1948, the Willamette River had overtopped its banks, inundating SP's Portland yard facilities journal-deep and demanding extreme caution from switching crews. C-8 #2722 and crew provided service where diesels dared not go.
(W. C. Whittaker: Menke collection)

Fresh from the Bayshore shops in early May 1955, C-8 #2725 stops for water on a cab hop at San Mateo. While the fireman attends to the spout, the engineer uses the main rod as a stepladder to adjust the inside Stephenson valve gear. Minor overspray on the lettering was characteristic of late-era steam paint jobs at both Bayshore and West Oakland. *(Jack Wirick: Menke collection)*

In the days before Felton was the home of a tourist railroad, the Santa Cruz area produced both lumber and sand. Leading the Olympia Sand Job at Felton on July 10, 1950 SP C-8 #2733 shuffles a selection of hopper cars, both loaded and empty. *(Charles Givens)*

The Colton roundhouse could not always provide a passenger engine for helper service on the eastbound heavyweight *Sunset Limited*, so on occasion a freight locomotive was called. About 1947, C-8 #2740 assists first #2 through Bryn Mawr, east of Colton. *(Howard Fairchild)*

Below, #2728 and an SD9 provide a rare mileage treat for fans on the Placerville Branch in the waning days of steam. *(Jim Wren: Rattenne collection)*

Alco-built #2750 illustrates the short smokestack with an exhaust deflector installed, but pulled back on its pivot. Although built as class C-8, the #2750 was converted to C-10 specifications in 1929. It was photographed on the afternoon of February 2, 1952 at Bakersfield and would serve two more years before being vacated in December 1954. *(Allan S. Youell)*

Above: Saving and swapping large tenders when a late model road locomotive was scrapped allowed SP to retire a few remote water facilities. A class 120-C-3 tender, normally used behind much larger locomotives, lends a purposeful look to C-9 #2777 at Fresno on August 2, 1955. *(Stan Kistler)*

Below, class C-9 #2828 was vacated three days before this picture was taken (at West Oakland) on October 9, 1954. Odd parts are stacked atop the boiler and blanks have been removed from the train number indicators. It still carries its oval "Harriman" builder's plate, but in a few months, the #2828 will be hauled around the Bay for scrapping in South San Francisco. *(Ray Whitaker)*

Mountainous branches often had snow problems, but SP could not afford to buy rotary plows for its secondary lines. Instead, on a seasonal basis, locally assigned locomotives were fitted with large, pilot-mounted wedge plows. These devices were held in reserve at major engine terminals for intallation during winter months. One such plow has been fitted to C-9 #2788 at Roseville in 1940. While they were at it, local mechanics added a foldaway exhaust deflector.

A decade spent working for the SP of Mexico as their #610 resulted in a few distinctive modifications for C-9 #2787. A feedwater heater installed on the fireman's side has displaced the air pump to the right side, while the Mexican preference for full ladders and matching handrails is still evident. Even a front toolbox had been added by the time of this photo, December 1948, at Eugene.
(Bert H. Warrd)

Routine steam-powered freight operations ran down in November 1956, after which most movements were to reposition locomotives or deliver them for stationary boiler service. On November 30, 1956, with water facilities already rare, the #2754 had run low on water and had to call on the San Leandro Fire Department for a refill. The engineer helps out in this end-of-an-era image.
(S. Romo, Jr.)

An unknown photographer was on the alert in Berkeley in May 1955 when C-10 #2831 passed by with a long eastbound extra freight. Even though the SP was rapidly dieselizing by 1955, small steam locomotives like this one were still being called for road jobs. A narrowed oil bunker on the tender and a footboard pilot suggest that #2831 spent most of its time at yard duties.
(Arnold Menke collection)

Class C-10 was SP's final iteration of the 1904 Common Standard 2-8-0. Nineteen of them, #2839-2857, were erected at Sacramento, Los Angeles and Ogden shops between 1917 and 1919 (World War I) when SP could not buy its own locomotive designs from commercial builders. Most of them used "swing boilers" swapped from other 2-8-0s, but at least the last five were ostensibly constructed from all-new parts. The #2842 (Sacramento, March 1918) rests at Pomona on February 2, 1941. *(Allan S. Youell)*

Opposite and below:
Artifacts of times past in San Francisco, a short train with a war-era boxcar caboose threads its way along the old main line diagonally across the City, west of Geneva Avenue. C-9 #2804 leads a short train on September 20, 1946; much of this valuable right-of-way later was recycled for transit and freeway development.
(Paul L. Henchy: Menke collection)

Class C-10 #2849 recently required testing of its power reverse gear, explaining the white wash on the levers. Compared to #2842 above, the #2849 (photographed on January 16, 1949 at Los Angeles) has received a small Pyle headlight and aswitcher pilot. In 1951, it would be leased to the SP de Mexico and sold later that year to become Ferrocarril del Pacifico #636. *(Allan S. Youell)*

SP had leased large 2-8-0s from subsidiary Cotton Belt ever since the late 1940s, even to the point of equipping some with train number indicators. In 1956, approaching final dieselization, SP decided to postpone a diesel switcher purchase for the Coast Division and instead lease eight 2-8-0s that still had flue time on their boilers. They became SP #2861-2868 in January 1956 and worked only a few months around the Bay Area before all steam switching operations were closed down. The #2863 passed through Los Angeles in March 1956 on its way north. *(Robert E. Smith)*

Considerably better performers than SP's "native" Common Standard 2-8-0s, the former Cotton Belt engines allowed SP to save interest charges on new diesels while providing photographers with the amazing opportunity to shoot "new" SP steam locomotives in 1956. Working in the Bayshore Yard that summer, the former SSW #773 (Baldwin 1923) would be idled for good the coming fall and be sold for scrap in 1960.

The railfan community was well aware that the recently arrived Cotton Belt 2-8-0s would have a short tenure on the SP, so the Pacific Coast Chapter of the Railway & Locomotive Historical Society sponsored this "hello and goodbye" fantrip in October 1956 to Modesto and back. The well-polished #2862 initially led the train, but it developed problems and was replaced by the grimy #2863; a water stop at Niles doubles as a photo opportunity. *(Fred A. Stindt)*

The Cotton Belt relied upon 2-8-0s as its primary freight power until 1930, taking development of this wheel arrangement farther than on the SP. Baldwin built 35 class K1 locomotives in 1920 and 1923 when the SP was buying 2-10-2s and 4-8-2s, but a few survived in good shape even beyond dieselization of the SSW. SP put them to work for a few months in the Bay Area, then closed out their leases and scrapped them locally. The #2861 (Cotton Belt #751, Baldwin, August 1920) pauses for a picture at Bayshore.
(Douglas S. Richter)

Class C-18 2-8-0 #3408 was a Baldwin 1903 graduate that began its days as El Paso & Southwestern #159. Smaller than a Common Standard 2-8-0, it was converted from coal to oil firing in 1929, but was to serve out its years on the Rio Grande Division, vacated from the roster in March 1950 at El Paso. It is pictured about 1948 with a semi-cylindrical ("whaleback") tender and a large pedestal headlight.

While boasting of "full dieselization," penurious SP managers chose to operate subsidiary Ferrocarril de Nacozari with steam locomotives #3400, 3406 and 3423, maintaining them in Mexico to remove them from ICC scrutiny. Finally, in 1959, Nacozari steam was supplanted by two 5200-series Baldwin diesels. For a brief period from 1957 to 1959, these three 2-8-0s shared honors with narrow gauge #9 as the only "SP" steam locomotives in regular service. An unkempt #3406 is pictured on March 4, 1957 at Douglas, Arizona.

(Harold F. Stewart)

Those are genuine New Mexico "black diamonds" in the coal bunker of the semi-whaleback tender trailing C-19 #3421. This 1904-vintage Baldwin 2-8-0 was inherited from the El Paso & Southwestern where it was formerly #272. When the EP&SW rebuilt this machine with superheated steam, the slide valves were replaced by Cotter valve chests and piston valves. Photographed in January 1946 at El Paso, the #3421 would not endure to be rebuilt as an oil burner; it would be vacated in November 1947.

(Allan S. Youell)

Coal smoke tints the New Mexico sky on May 11, 1940 as westbound mixed train #975 departs Tucumcari for Dawson. Class C-18 #3401, a 1902-vintage Baldwin, faces an uphill climb on its 131-mile, all day task. Mines on this branch were to provide fuel for SP steam locomotives until 1950; after that, SP began converting its coal-fueled steam locomotives to oil, as well as bringing in new diesels. *(Richard H. Kindig: Menke collection)*

When SP took over the El Paso & Southwestern in 1924, that road's large coalburning 2-8-0s were soon transferred to the SP de Mexico and converted to oil fuel and eventually received new valves, pistons and running gear. SP had plenty of 2-8-0s of its own, so #3440-3449 were given SPdeM numbers and worked there until the early 1940s. Upon outbreak of World War II, they were called back to the Pacific Lines and re-lettered for SP. The #3449, fresh out of the shop in 1942, breaks in on the San Pedro line along Alameda Street south of Los Angeles. Light dispersion shields have been installed on the back-up headlight and marker lights. *(H. L. Kelso)*

Unlike SP's Common Standard 2-8-0s, few (if any) of the larger 3400s ever were involved in tender swaps; most pictures show them with EP&SW-style rectangular tenders with fuel bunkers modified to accept oil tanks. In 1929-30, they were sent to the SP de Mexico shops at Empalme, Sonora, for conversion to oil fuel and addition of feedwater heaters, then operated in Mexico with SPdeM numbers until SP ran short of locomotives. On October 22, 1941, this C-26 was renumbered from SPdeM #653 to SP #3443 and is pictured a month later at Los Angeles. In 1945, it will be sent back to the SPdeM, retaining its SP number. *(G. M. Best)*

The eighteen class C-28 locomotives in SP's 3452-3469 series carried four or even five different numbers in their four decades of service: two for the EP&SW, one for the SPdeM and one for the SP. Sixteen were sold in 1951 to the Ferrocarril del Pacifico, where they received a final set of numbers. Fresh from repaint and renumbering in July 1941, the #3460 waits for clearance at Lomita Park, near San Bruno.
(Fred A. Stindt)

Steamed up and ready to go at Los Angeles in February 1942, #3464 has yet to be fitted with an "anti-dispersion" headlight shroud. While otherwise similar to the #3460 above, the boiler walkway was installed completely differently. Soon after the war was over, this engine returned on lease to Mexico, retaining its SP number. In December 1951, SP #3464 will become Ferrocarril del Pacifico #662.
(G. M. Best)

SP's Pacific Lines did not routinely install front ladders on older locomotives, but both the EP&SW and the SP of Mexico adopted this feature. Class C-26 #3444, originally EP&SW #64 (Baldwin 1906), spent most of its SP years south of the border as SPdeM #668. Briefly repatriated in 1941 as SP #3444, it would be leased back to Mexico in 1948 and then sold to the Ferrocarril del Pacifico as their #644. Amazingly, it went on from there to become Nacionales de Mexico 2nd #1803, where it was finally retired in 1960. It was photographed in September 1948 at Los Angeles.

In 1946, SP leased Cotton Belt 2-8-0s #554, 560, 562-3, 569, 574 and 577 for use on the Tucson and Rio Grande Divisions. They were given aluminum-painted smokebox doors and train number indicators, but were never renumbered to the SP. The #562 and 563 were scrapped by the SP, but the others were returned to their owner in 1949. In 1951, SSW #577 (photographed at Tucson on November 8, 1947) would become Ferrocarril del Pacifico #680. *(R. J. Foster)*

Class C-28 2-8-0s #3452-3469 came from the El Paso & Southwestern. Alco-built in 1907, they boasted Walschaerts valve gear, which their contemporaries on the SP would never receive. They were rebuilt 1929-30 as oil burners on the SPdeM; original 25x30-inch cylinders were reduced to 23x30 in subsequent shoppings. In 1940-42, they returned to the Pacific Lines, with most being sold to the Ferrocarril del Pacifico in 1951. The #3463 appears at Los Angeles on November 7, 1948. *(Stan Kistler)*

On April 25, 1948, the Northern California Railway Club sponsored an excursion to Davenport, geographically northwest of Santa Cruz, on trackage off-limits to large passenger locomotives. This photo location is near the station point of "Wilder Spur." The road engine was ten-wheeler #2374, doubleheaded with 4-8-0 #2915; instead of the usual baggage car behind the locomotives for those seeking aural thrills, the SP provided an open-end observation abutting #2374's tender. *(R. Willey)*

Chapter Seven
Twelve-Wheelers: 4-8-0

The oldest road locomotives still operating on the SP during the 1940-57 era were the Twelve-wheelers in the 2900 series. Senior among them were five TW-4 engines, #2926-2927 and 2929-2931, built by Cooke in 1883. Next in age were the dozen TW-3 and five TW-2 locomotives delivered in 1892-93 by Schenectady, followed by the single TW-6 (#2957).

These twenty-three 4-8-0 locomotives were still in use (and well maintained) into the late 1940s. They weighed between 65 and 70 tons loaded, tipping the scales between 51 and 60 tons on their eight driving wheels. The lightest of them 27,750 pounds and the heaviest 30,075 pounds per driving axle—generating between 83 and 88 per cent of the tractive effort of an M-6 2-6-0 (SP's favorite "little engine"), while loading each driving axle only 55-60% of the weight applied by each of the three M-6 driving axles.

Many branchlines on the Portland Division had bridges with a maximum load rating of twenty tons (40,000 pounds) per driving axle. As a result, the Operating Department was forced to either retain old locomotives that fell within those weight limits or lobby senior management to buy new locomotives that fit existing weight limitations. Tipping the decision in favor of new lightweight locomotives was the fact that SP could always use the locomotives elsewhere, but money spent on infrastructure would be lost if and when the branch line lost traffic (*i.e.*, logged out) and was abandoned. The bottom line was that it was cheaper to run old steam or small new diesels over unimproved track than it was to upgrade the track itself.

As soon as funding was available for 70-ton GE locomotives (at $73,000 each), their lower cost of operation carried the day. In fact, SP purchased ten 70-ton GE diesels in 1949 and immediately scrapped thirteen twelve-wheelers.

The two classes of heavier, more modern 4-8-0s (TW-1, #2900-2913 and TW-8, #2914-2923) still weighed less per each driving axle than either a Common Standard 2-6-0 or a C-9, allowing them to serve trackage where tractive effort-intensive switching was required, but where stringent axle load limits applied. One TW-8 (#2922) was leased to the Ogden Union Railway & Depot Company, while five more TW-8s and two TW-1s worked out of Watsonville Junction on the Coast Division. The Portland Division maintained four TW-8s as well as most of the old 4-8-0s. On the San Joaquin Division, TW-8 #2914 was already on its station at Bakersfield and would remain there until retirement. Finally, the Rio Grande Division maintained the #2933, 2935 and 2942 .

These useful machines finally were replaced by diesels, almost all of the 4-8-0s leaving the roster between 1947 and 1951. Light Alco diesels replaced TW-1 and TW-8 engines on Salinas packing shed trackage, while the #2937 worked as a "loaner" for various assignments in the Bay Area. The #2914 worked on the Sunset Railway west of Bakersfield until early 1955; it was donated to the city later that year and remains on display there in 2010.

The local serving Lewis Springs (Fort Huachuca) from Benson, Arizona, was train 940; on December 9, 1946 this service rated nothing but an auxiliary water car for the locomotive and a weary wooden combine. Class TW-3 #2935 began its years of service in July 1892 as a compound; it was rebuilt to simple in December 1909. Not long after this photo, the #2935 was damaged in a wreck and the hulk was shipped to Oregon as a parts supply, with the remains scrapped November 13, 1947 at Brooklyn. *(A. C. Phelps: Dunscomb collection)*

Above: Pressed into mainline service early in World War II, #2904 has a hood over its headlight and on April 18, 1942 is in charge of a 33-car freight at Takoma Street in Brooklyn, southeast of downtown Portland. This twelve-wheeler will serve until late 1949, when GE 70-ton diesels will arrive. *(Arnold Menke collection)*

Below: Branches near Bakersfield were the final redoubt for 4-8-0 operations; class TW-8 #2914 used an auxiliary water tender to venture into the southern San Joaquin Valley badlands where there were no permanent water tanks. Photographed on March 19, 1955, the #2914 will be vacated that May for permanent display at Bakersfield. *(John E. Shaw, Jr.)*

Nicely painted and lettered at the Brooklyn roundhouse on September 3, 1948, 53-year-old TW-1 #2900 appears to be ready for another fifty years of service. Replacement diesels, however, are already in SP's budget and #2900 has only about two more years left. The rectangular tender is a class 90-R-7, commonly seen trailing 0-8-0 switchers. A blowdown spreader muffler is installed between the second and third drivers and a power reverse has been added above the rear pilot wheel. *(Guy L. Dunscomb)*

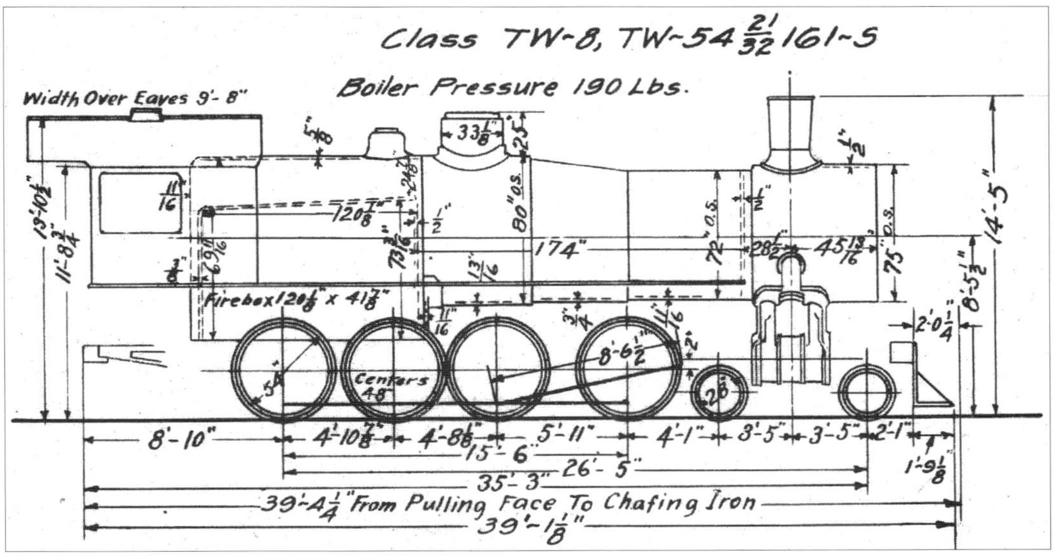

Twelve-wheelers were preferred for switching light rail around Salinas and Watsonville vegetable packing sheds. At Salinas on November 24, 1946, TW-8 #2915 has been fitted with a backward-facing power reverse and an aluminum-painted smokebox door, but the 7000-gallon "clear vision" Vanderbilt tender (with the oil bunker trimmed back to provide better sight lines during switching) still carries "Southern Pacific Lines" lettering. Re-locating the headlight to the center of the smokebox door was a simple stroke of modernization that brought this TW-8 into modern times. *(W. C. Whittaker)*

The highest-numbered TW-1 was built in July 1895 by Schenectady (years before "Alco" was organized) and placed in service on August 3, 1895. Pictured at Ogden on August 14, 1947, #2913 features slide valves (no SP TW-1 was ever rebuilt to superheat) and archaic smokebox-top headlight placement. SP routinely assigned one or two 4-8-0s to the Salt Lake Division, often for use by the Ogden Union Railway & Depot Company. *(W. H. Thrall; Shaw collection)*

Del Monte, one station stop away from Monterey, shared the region's mild climate. An overcast April 28, 1940 finds superheated TW-8 #2915 leading a short extra past the passenger platform. At this point in time, the 1898-vintage locomotive still carried a large "Golden Glow" headlight and a class 73-SC-1 whaleback tender (built by Baldwin) in contrast to its more modern appearance on the previous page. *(W. C. Whittaker: Menke collection)*

The #2934 was built by Schenectady in 1892 as a two-cylinder cross compound; SP rebuilt it to simple in April 1910. Long a stalwart of light-rail Oregon branchlines, this class TW-3 is pictured on November 12, 1946 at the SP's Brooklyn Roundhouse, across the Willamette River from Portland. Despite modern lettering, the #2934 still carries an old pedestal headlight and spoked pilot wheels. It will be vacated from the roster in May 1949. *(Leo J. Munson collection)*

Still carrying its Schenectady builder's plate, TW-3 #2940 is pictured on June 18, 1949 at the Albany engine facility some eighty miles south of Portland. Modernity has arrived in the form of solid pilot wheels, a Pyle headlight, large tender lettering, and switcher footboards. GE diesels are reporting for service; this useful little machine has only a few months left of active duty. It will be vacated at Brooklyn in January 1950 and scrapped there in August. *(Dick Wolf)*

The only example of class TW-6 to survive into the 1940 era was #2957, the highest-numbered of latter-day 4-8-0s. After many rebuildings, there was little to distinguish one twelve-wheeler class from another, but original #2954-2957 were always simple engines—they never ran as compounds. On the traditional ready track at the Brooklyn roundhouse on August 2, 1948, #2957 has been painted and polished for a final tour of duty; it will be stricken from the roster in June 1949.
(D. H. Roberts: Dunscomb collection)

The last "old style" 4-8-0 to grace the roster was TW-3 #2937; SP found a need for it for a few months (August 1952 to May 1953) on light-rail segments of the Northwestern Pacific north of Eureka, but in November 1953, this venerable machine was finally written off. Pictured in brilliant sunlight at Brooklyn on November 17, 1947, #2938 was assigned this modern Vanderbilt tender in November 1938 and kept it until retirement. *(Don H. Roberts)*

The oldest road engine on the Pacific Lines during the period addressed by this book, TW-4 #2926 was a Cooke product dating from December 1882, built as Southern Pacific of California #52. Weighing just 72 tons with a full boiler, it spread 52 tons over its eight drivers, allowing it access to the flimsy bridges and light rail common on Oregon branch trackage. At Eugene on September 1, 1948, this 66-year-old features latter-day lettering and a Pyle headlight and is ready for another day's work. It had six months left to serve at this moment. *(Guy L. Dunscomb)*

Near the present site of one of the world's busiest McDonald's Restaurants (on Interstate #5 near Buttonwillow, California), TW-8 #2914 darkens the Kern County sky with a pillar of oil smoke on its way west from Bakersfield early in 1955. Once this engine is retired a few months after this picture, the era of the twelve wheeler on the Southern Pacific will be over. *(John E. Shaw, Jr.)*

Below: At-work pictures taken in the 1950s of 4-8-0s (other than the #2914) are scarce; as the oldest steam engines on the SP roster, they faded away as diesels became more common. It's very late in the #2938's working career—May 1951—but the *Wendling Turn* at Springfield, Oregon, was still powered by a twelve-wheeler. The Eugene-Springfield area was home to numerous lumber mills that required virtually constant service, hence the longevity of this nearly sixty-year-old locomotive. *(Bert H. Ward)*

In modern times, the only all-year shipper on Tehachapi Pass was the large cement works at Monolith, which had its own narrow gauge railroad. Empty boxcars for bagged cement and covered hoppers to haul the bulk product were delivered by locals based at Mojave. About 1950, an eastbound peddler (extra 3222 east) works downhill from Monolith and Tehachapi near Warren, on its way home to Mojave.

Chapter Eight
Mikado: 2-8-2

By the closing years of Union Pacific control, the SP had an oversupply of Common Standard locomotives, especially 2-8-0s. Locomotive design continued to evolve, however; SP ordered its first 2-8-2s, really just an evolution of the class C-9 2-8-0. Baldwin built Mk-2s #3200-3215 in 1911, complete with 57-inch drivers and long-stroke cylinders—but utilizing an early application of Walschaerts valve gear.

Within two years SP designed a 2-8-2 with re-aligned piston valves that allowed for less complex valve gear; the Mk-4 put these features to work, retaining the same basic specifications and performance. Baldwin built Mk-4s #3216-3235 in 1913. Later that year, in a "last act" project by the Harriman Associated Lines, a dual-service 2-8-2 was debuted, the Mk-5, with 63-inch drivers and short-stroke 26x28 cylinders. The Harriman roads used these machines on mountain passenger trains as well as on freights; SP obtained 14 from Baldwin in 1913 (#3236-3249) and then twenty virtually identical Mk-6s from Lima in 1915: #3250-3269. (Incidentally, these Lima machines marked SP's first order for road engines from the Ohio builder.)

Baldwin Mk-5 #3270 was built "on spec" for the Panama-Pacific International Exposition in San Francisco—with the obvious expectation of selling it to the SP once the Expo had closed. In 1924, five Mk-4 locomotives were added when the Arizona Eastern was taken over; it was decided to number them in series with SP's own Mk-4s as second #3236-3240, displacing the SP engines to unoccupied #3271-3275. Also arriving in 1924 from the AE were 1917-vintage Lima Mk-5s #3276-3277.

A major addition to the roster came as a result of SP's 1924 lease of the El Paso & Southwestern. Essentially identical coal-fired classes Mk-7, -8 and –9 locomotives #3300-3324 (Alco-Schenectady 1913-1920) were all added at once. EP&SW standards were far different from those of the SP; that road believed in low boiler pressure and very large cylinder diameter, while SP practice was significantly less conservative. Over the next two decades, SP shop forces labored mightily to bring these machines into line—most importantly reducing cylinder diameter, raising boiler pressure and converting them to oil fuel. They were well distributed over the Pacific Lines by the period discussed in this book.

In 1935, SP foreclosed on some loans it had made to the connecting Minarets & Western, obtaining in the process two Alco-Brooks 2-8-2s (Mk-10, #3295-3296) that compared favorably—except for their 51-inch drivers—with the home-grown product. Then in 1940, when new locomotives were already difficult to obtain, SP bought a duo of small logging 2-8-2s, putting them to work as class Mk-11, #3297-3298. They had been built in 1914 by Alco's Brooks Works.

(Continued on page 113)

The sixteen class Mk-2 Mikados, built by Baldwin in 1911, were SP's first non-articulated locomotives built with Walschaert valve gear, featuring small valve chests similar in profile to those on the 2-8-0s, placing the valve stem far inboard of the radius rod and combination lever. To compensate, the builder ran two long parallel rods from the cylinders back to the crosshead guide support to act as valve stem crosshead guides. A better solution was to make the valves larger in diameter and move their centerline outboard to place the valve stem in line with the valve gear–with a short guide supported by the back valve chest head. This improvement was used on later 2-8-2s as well as rebuilt 4-4-2s, 4-6-2s and those few Mk-2s rebuilt as Mk-4s. The #3209 models the original arrangement on July 27, 1949 at Los Angeles. *(Allan S. Youell)*

Extra 3207 east, the *San Fernando Local*, passes through Sun Valley north of Los Angeles on August 14, 1949 as it returns to Taylor Yard in Los Angeles. Fitted with a clamshell smoke deflector and switcher footboards, Mk-2 #3207 is performing a job that will later be assumed by Baldwin roadswitchers. *(Walter H. Thrall: Shaw collection)*

Common Standard Mk-2 #3204 is pictured on November 6, 1949 at Dunsmuir, illustrating the original cylinder saddle with its small valves. This arrangement limited performance; by 1913 SP had seen the light and ordered its new Mk-4 locomotives with considerably larger valves. Mk-2s #3203 and 3208 (across the page), among others, were rebuilt in the 1920s with larger valves. The #3204, despite its apparent handicap, remained useful until retirement in late 1951.

Significantly upgraded by the SP since leaving the Baldwin Works in 1911, #3208 was rebuilt in 1929 from class Mk-2 to Mk-4. A sport cab, front frame brace and clamshell smoke deflector add to the more modern look—although SP never went so far as to rebuild 2-8-2s with Delta trailing trucks or boosters. Pictured at Indio on December 9, 1951, the #3208 will soon migrate to northern California and work on the Western Division. *(Stan Kistler)*

SP railroaders must have looked with envy at "poor neighbor" Western Pacific's 327 and 332 class Mikados, with their trailing-truck boosters and 15,000-gallon tenders. In fact, SP never modernized any 2-8-2 with a Delta trailing truck. Power planning at the SP was dominated by an overabundance of Harriman-era 2-6-0s and 2-8-0s, to the point where "fast freight" became a meaningless expression on the SP.

Future historians may discover an explanation for the early demise of so many Mk-2 and Mk-4 engines. The #3200 was wrecked during hard times and vacated in 1934, but the #3210, 3220, 3223, 3231 and 3233 never made it through the Depression years. Some SP 2-8-2s were rebuilt with "heavy frames," suggesting that there was a fundamental engineering flaw in the early machines. Otherwise, all the 2-8-2s remained active until the 1950s, when they were displaced by diesels.

The #3203 illustrates the fireman's side of the rebuilt Mk-4, photographed on March 17, 1956 at Bayshore. This engine was rebuilt at SP's Ogden Shops in March 1928. Low light emphasizes the modern cylinder saddle featuring large outward-sloping steam valves. As was common SP practice, large tenders were rarely scrapped—just passed down to smaller locomotives. This 12,000-gallon semi-cylindrical tender once could have been seen behind a Mallet or a 2-8-4. *(Ray Whitaker)*

Both the southern and northern urban areas of California required huge quantities of sand and gravel for construction. In the Bay Area, this need was met by a vast open-pit quarry complex in the Livermore Valley between Pleasanton and Livermore. The station point of Radum probably originated more gross tonnage than any other location in northern California—most of it destined for short hauls. In April 1954, the *Radum Sand Job* meets eastbound freight #408 (with helper 4-8-4 #4461) at Radum. *(Charles Givens)*

Baldwin delivered class Mk-4, #3216-3235, in June 1913. They were built with larger valves and less restrictive steam passages, but otherwise followed the Mk-2 pattern, including the Hodges trailing truck. Many were rebuilt with sport cabs and front frame braces, but in general they required little modification over the years. By April 19, 1956, not much time remains for #3224 at Fresno; it will be vacated from the roster in less than four months. *(Ray Whitaker)*

Above: Extra 3237 west rolls into Mojave in May 1950. This long train dropped off its helper at Tehachapi Summit, but picked up eight ponderous carloads of bulk cement at Monolith for this final lap into Mojave. The Mikado features a clamshell smoke deflector and a switcher pilot finished off with a corrugated insert. Second #3237 was the former Arizona Eastern #901, absorbed into the Pacific Lines roster in 1924. Between 1937 and 1941, it was leased to the SP de Mexico, but returned to the SP before World War II and would serve until vacated in May 1953. *(Arnold Menke collection)*

Right:
Two Mikados, both photographed at Los Angeles within a short timespan, illustrate the visual impact of a long tender and a sport cab, as opposed to a short rectangular tender and square cab. Mk-4 #3235 is pictured on January 16, 1949. Dimensionally similar Mk-2 #3222 posed on November 16, 1946. Although the smokebox door has been painted aluminum, the rectangular tender (a pressurized class 90-R-1 from one of the original 2-8-8-2s) still carries "Southern Pacific Lines" lettering. *(Both: Allan S. Youell)*

Mikados were favored for gravel trains originating in the Livermore-Pleasanton area. SP #3240, actually the second 2-8-2 to carry that number, began its days in 1913 on subsidiary Arizona Eastern, where it was #904. This Mk-4 was renumbered second #3240 in 1924 (first #3240 became #3275 to clear the number) and spent many years on lease to another subsidiary—the SP of Mexico. It wasn't renumbered there, but the SPdeM favored a full "staircase" up to the running boards, a feature that remained on #3240 for the rest of its career.
(Jim Harrison collection)

The Mk-5 class, originally #3236-3249, was designed for true "dual service," not just slow freight. Sixty-three-inch drivers replaced the fifty-seven-inchers used on the Mk-2 and Mk-4, while cylinders were larger in diameter but with a shorter stroke. Ready for service, the Mk-5 weighed 140 tons, while its predecessor 2-8-2s scaled 123 tons. Both SP and Union Pacific used Mk-5s on passenger trains as well as freights. That tender is class 120-C-7 and was formerly used behind a coal-fired 2-10-2 in New Mexico. At Bayshore on July 11, 1955, #3247 will serve only a few more months. *(D. S. Richter)*

At Aztec, Arizona, blasting up the 1% grade east of Yuma, the fourth section of Tucson Division train #842 is powered by class Mk-5 Mikado #3242. It's January 21, 1947 and SP has new F3 diesels on order—but their arrival is still months in the future. Meanwhile, eight Maintenance-of-Way water cars have been added to #3242's burden, apparently consigned to SP's desert outposts. *(R. P. Middlebrook: Menke collection)*

An aluminum-painted smokebox and large SOUTHERN PACIFIC tender lettering are a few months in the future on October 8, 1945 for #3248 at Tracy—but World War II came to an end just a month before this picture; headlight dispersion shields have already been consigned to the scrap heap. The air pump has been displaced to engineer's side due to installation of a feedwater heater. The #3248 is one of a number of Mikados rebuilt with what SP described as "heavy frames," signified by the reinforcing girder across the pilot beam. *(David J. Welch: Menke collection)*

Twenty class Mk-6 Mikados, #3250-3269, were built in late 1914 by Lima; they were close copies of the previous Mk-5 design, providing a foot in SP's door for sales by the Ohio locomotive builder. Freights leaving Bakersfield for Fresno frequently required a 2-10-2 to handle the tonnage, but on October 12, 1951, Mikado #3261 is sufficient for westbound #785 as it departs the Bakersfield yard.

(Allan S. Youell)

Both Mk-6 #3269 and Mk-5 #3270 boasted atypical (at least on the Southern Pacific) installations of Elesco feedwater heaters on their smokebox brows. On July 21, 1941, begrimed #3269 has drawn the helper assignment for today's *Noon Daylight,* powered by a very new GS-4. Train #97 is pictured at Hathaway, geographically north of San Luis Obispo, on the ramparts of Cuesta Grade.

(Lewis Harris: Menke collection)

Considerably more capable than SP's ubiquitous Harriman 2-8-0s, the Mikados could complete a given job in less time and at less fuel expense. Even in the final months of steam operation on the Coast and Western Divisions, 2-8-2s were found at work on locals—jobs that would be inherited by Baldwin and Alco roadswitchers. The obscure station point of Simla, on the Los Altos Branch, was located near milepost 39.7; a spur diverged here to the Kaiser Permanente Cement plant. Lima Mk-6 #3251 was working the spur on June 18, 1956. *(Charles Givens)*

SP didn't have many customers in the arid lands surrounding Mojave, but those few generated tonnage that required at least a Mikado to manage; as a result, there always seemed to be a few 2-8-2s stationed there. The large twelve-wheel tender and sport cab featured on Lima 2-8-2 #3266 are illuminated in perfect low light at Mojave on March 18, 1951. This particular Mk-6 would serve another seventeen months before being vacated at Bakersfield in October 1952. *(Allan S. Youell)*

The Panama-Pacific International Exposition was the equivalent to a "world's fair," held in San Francisco to celebrate the opening of the Panama Canal. The Baldwin Locomotive Works exhibited at least three new locomotives there, all of them Mikados. In October 1914, apparently on speculation, they built an Mk-5, lettering it for the Southern Pacific. SP bought it in December 1915 as their #3270. In 1920, it received a test installation of an Elesco feedwater heater—which remained with this locomotive the rest of its days. Less than three months after this image at Colton on February 22, 1953, the #3270 would be vacated and scrapped. *(Allan S. Youell)*

Three gondolas, a baggage-RPO and a coach make up mixed train #605 (originating at Hazen, Nevada) on September 2, 1946, at the faraway outpost of Wabuska, Nevada—then the SP's junction with the Nevada Copper Belt. Built in 1913 as SP #3239 and renumbered in 1924 to #3274, class Mk-5 Mikado #3274 has yet to receive its postwar paint scheme, but boasts a feedwater heater installed at Ogden Shops only six months prior to this photo. *(W. C. Whittaker: Tom Moungovan collection)*

Slow passenger trains in mountainous territory were often candidates for Mikado power, as demonstrated on #327, the Dunsmuir-Grants Pass (Siskiyou Line) local passenger train once known as the *Shasta*. Mikado #3273 makes its station stop at Hilt, California, in June 1951, the first station stop inside the California border. The train will be discontinued in February 1952, less than a year after this photo. *(Charles Givens)*

SP foreclosed on a bankrupt connecting shortline, the Minarets & Western, putting its equipment up for sale. Two M&W 125-ton 2-8-2s, smaller than any SP design, were kept by the SP, becoming #3295-3296 in 1935. With 51-inch drivers, they were best utilized as yard switchers; a job the former M&W #102 (built by Alco-Brooks in 1923) performed admirably at Dunsmuir for many years. Even though SP added its own features, the #3295 (at its home yard on August 24, 1952) retained a distinctive "look." *(Richard V. Dodge: Menke collection)*

At work in July 1952 on a local from Coos Bay to Myrtle Point (where Coos Bay Lumber Company trackage began), Alco Mk-11 #3297 pauses for a picture at the outpost of Johnson, Oregon. When the SP bought #3297-3298 in 1943, they were initially assigned to the (SP) Brooklyn—(UP) Albina transfer run, but after a few years were reassigned to switching at Eugene. Finally, they were moved to Coos Bay for local work. Time ran out in 1953-54 for these two small Alcos. *(Bert H. Ward)*

As long as SP ran regular trains with steam locomotives, those complex machines were well maintained and ready for routine work. Cosmetic standards slipped a bit, but the Mechanical Department remained proud of its charges. Waiting for its crew on the Bayshore ready track on June 1, 1956, 43-year-old Mk-5 #3275 has many good miles left, although dieselization will catch up with it in two more months. The class 120-SC-1 whaleback tender features a raised oil tank. *(D. S. Richter)*

High water spilling over from the Willamette River impairs switching operations in Portland on June 8, 1948. Only a little more secure than a diesel-electric in these circumstances, #3298's tender is journal-box deep in floodwater. The secondhand Mk-10 and Mk-11 engines rolled on tiny (for a railroad-size Mikado) 51-inch drivers, rendering the machinery all the more vulnerable to water damage. This location appears to be at NW Thurman Street, not far from where today's Interstate 405 crosses the river. *(W. C. Whittaker: Menke collection)*

In 1943, SP was desperate for locomotives. War-time traffic kept increasing and government agencies were doling out only small numbers of new locomotives. When an equipment dealer offered two 112-ton logging-style 1914-vintage Mikados, SP snapped them up as class Mk-11, #3297-3298. Almost as powerful as a C-10 and much easier on the rails, these two little Alcos served a decade on the Portland Division. Working a Union Pacific transfer at Albina, Oregon, in February 1950, Mk-11 #3298 was built by Alco's Brooks Works for the Chicago & Illinois Midland. *(Don H. Roberts)*

Alco built this 2-8-2 in 1913 as El Paso & Southwestern #366, a coalburner with a small rectangular tender. In 1924 it became SP Mk-7 #3303. Extensively rebuilt in its SP days, it was converted to oil fuel in 1936, eventually receiving this 12,000-gallon "whaleback" semi-cylindrical tender (class 120-SC-3) that was originally built for Mallet #4004. It was photographed on November 22, 1953 at Tracy; the corrugated steel pilot is a latter-day innovation. *(D. S. Richter)*

Ready for its bath at Los Angeles in December 1947, #3304 contrasts with the slightly more modern look of #3303 on page 124 ; it still carries an old-style boiler-tube pilot and retains spun steel covers over the valve and cylinder heads. Removal of the latter feature was a general postwar project. *(G. M. Best)*

Opposite, below:
On July 14, 1953, Mk-8 #3311 leads westbound freight #429 from Tracy across Altamont Pass, a mile east of Altamont station. This train fulfills the traditional role of providing empties for Livermore Valley gravel pits near Pleasanton, the largest single tonnage generator on the Western Division. The smoke plume of the helper engine appears far to the rear, not far from WP tunnel #3. *(Herman Friedrich)*

Two dozen Alco Mikados, delivered to the El Paso & Southwestern in three batches between 1913 and 1920, were designed to standards foreign to SP practice. They were coal burners utilizing low boiler pressure (only 176 pounds *vs.* common practice of 200-225 at the time), but with huge 29" x 30" cylinders using great volumes of steam to "earn back" a respectible tractive effort figure. SP inherited them in 1924 as #3300-3324. Two-thirds were changed to oil fuel in 1929-30, before the Depression brought shop layoffs. Between 1938 and 1942, all but four were rebuilt with the smaller 27" x 30" cylinders and increased boiler pressure illustrated in this diagram. After the war, the stragglers were brought into conformity.

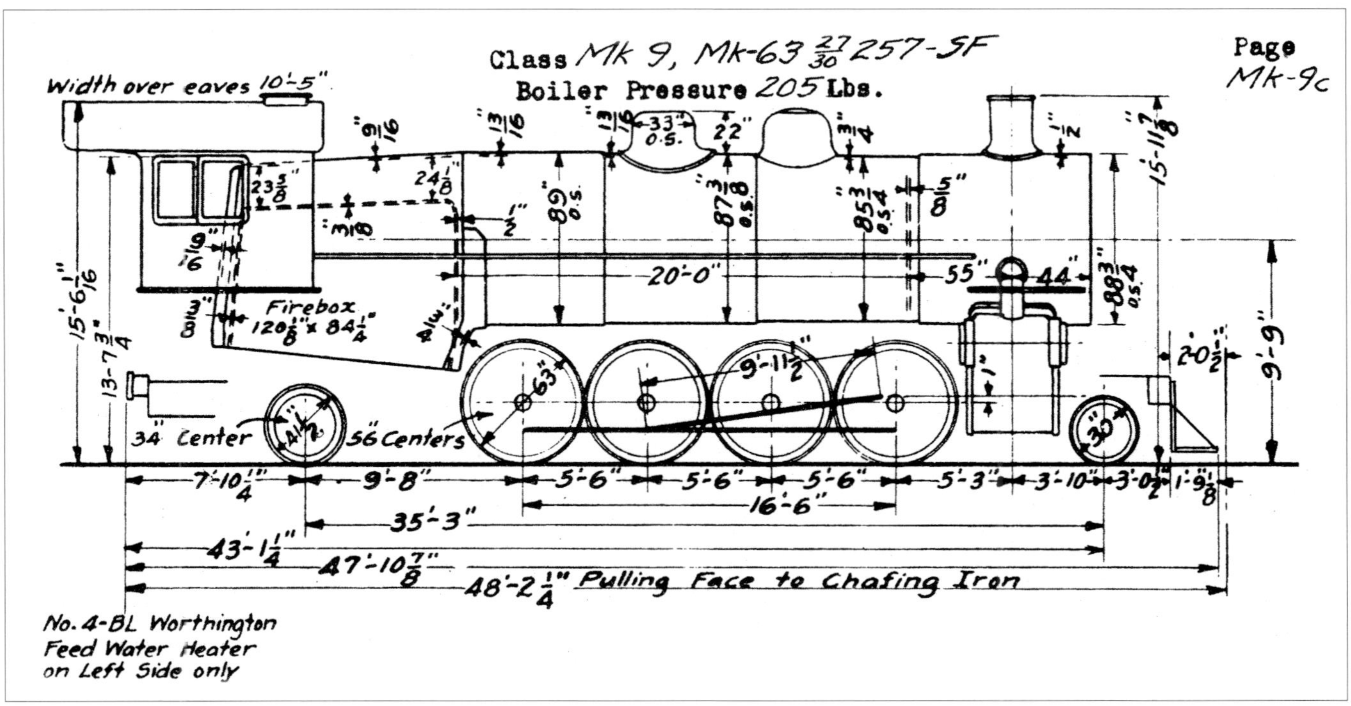

Almost ready to roll (but still without a train or extra number in the indicators), Mk-9 #3321 will be earning revenue soon. By November 1947, when this picture was taken at Tucson, SP had caught up with its deferred war-time maintenance; locomotives were sent out clean and modernization (or money-saving) measures appplied. The #3321 boasts a visorless Pyle headlight, dual air pumps and a corrugated steel pilot; in a few years it will be assigned a 12,000-gallon semi-cylindrical tender.

(Leo Munson collection)

Still carrying its pre-war lettering and paint and still equipped with a 9,000-gallon rectangular tender, the #3311 is overdue for modernization, but so was virtually every SP steam locomotive right after World War II. Leading an eastbound extra freight out of Oakland past El Cerrito in August 1946, #3311 has more than seven years of active service ahead. *(Stan Kistler: Menke collection)*

The former EP&SW 2-8-2s, #3300-3324, had to wait until larger locomotives (especially the 4-6-6-2s and 2-8-8-2s) were retired after World War II before they could inherit their 12,000-gallon semi-cylindrical tenders. Previously they used their original rectangular 9,000-gallon Alco tenders with the coal bunker converted to oil, as illustrated by the #3312 on September 2, 1940 at Roseville.

(Allan S. Youell)

Pictured about 1950 at Niles, Mk-9 #3323 carries a more modern Pyle headlight and leads a hand-me-down semicylindrical (whaleback) tender. According to the date stenciled on the cylinder heads, the #3323 was last serviced in December 1949 at Bayshore, probably the date when it lost its cylinder head covers. The light-colored rectangular device near the rails behind the last driver was SP's blow-down spreader, designed to direct boiler sludge onto the ballast, rather than out on the right-of-way.

(Strapac collection)

In 1945, the Southern Pacific bought ten secondhand coal-fired 2-8-4 locomotives from the Boston & Maine and put them to work in New Mexico. SP's obligation to use coal as locomotive fuel on the Rio Grande Division came to a close in late 1949; Berkshires #3500-3509 (which still had useful miles to go) were converted to oil fuel and transferred to assignments in central California—but they served only a few months there before being retired. Extra 3502 west charges out of Antelope Yard on September 9, 1950.

(W. C. Whittaker)

Chapter Nine
Berkshires: 2-8-4

In order to meet World War II traffic exigencies on the Rio Grande Division in a time when new locomotives were simply unobtainable, SP officers took a step their predecessors in World War I never imagined—buying secondhand eastern steam locomotives to keep the trains moving. "Not invented here" definitely applied to the ten 2-8-4 locomotives SP purchased in July 1945 at a cost of about $41,000 each. They were formerly owned by the Boston & Maine; 1928-vintage Lima coalburners that had previously operated in a soft-water region (hence their "closed" Coffin feedwater heaters).

Class B-1 entered service in mid-August 1945. Originally they were equipped with Baker valve gear, but the B&M had converted their #4000 (SP #3500) to Walschaerts and installed multiple-bearing crossheads. Among the spare parts SP received with the locomotives were valve gear and crosshead conversion kits to modify five more 2-8-4s; El Paso shops quickly performed this task on #3501-3505.

SP's contract for coal fuel on the Rio Grande and Tucson Divisions was due to run out in 1949-50, and dieselization loomed on the far horizon, giving SP the incentive to convert coal-fired steam locomotives that still had useful miles to oil fuel. In late 1949 and early 1950, the El Paso Shops converted B-1s #3500-3509 to oil fuel and removed their booster engines.

Typical of Lima 2-8-4s of this vintage, they were built with articulated engine frames, rather than a simple trailing truck. Their unique rear drawbar arrangement meant that SP could not just couple up its standard Vanderbilt tenders and send them out—what was done instead was to salvage the original Lima tender frames and trucks and mount semi-cylindrical tanks from retired articulated locomotives on the old frames. This arrangement coincidentally retained the 12,000 gallon water capacity. The #3503-3509 also lost their distinctive feedwater heaters during this shop visit.

They were immediately transferred to the San Joaquin Division, before feedwater heaters had been removed from #3500 and 3501 (Roseville and Sacramento shop forces quickly removed this feature from both). By all accounts they performed useful service, but they were "foreigners" and debuted on the eve of dieselization. Despite their shop modifications, the #3500-3509 served for just a few months in central California before new diesels arrived to displace them; vacate dates ranged from October 1950 to July 1951.

The original B&M series included twenty engines, #4000-4019. In an interesting parallel, the Santa Fe took seven of these odd machines as well, numbering them 4193-4199 and dramatically rebuilding one to Santa Fe standards—apparently just to prove that it could be done.

Less than six months after arriving for service, the former Boston & Maine #4003 had received only a splash of new paint and lettering, a blowdown spreader and train number indicators when photographed at the El Paso roundhouse on an overcast January 12, 1946. Coal fired, with Baker valve gear, a trailing-truck booster and an odd Coffin feedwater heater, 2-8-4 #3503 has almost nothing in common with accepted Southern Pacific practice—but SP really needed locomotives in 1945-46. *(Guy L. Dunscomb)*

Once World War II was over, the El Paso shops had the time to address on steam locomotive "cosmetic aspects." Still reflecting B&M practices with coal fuel, Coffin feedwater heater and a booster, the #3504 has been repainted and given large tender lettering on its rather small 12,000 gallon rectangular tender. Subsequently rebuilt by the SP with Walschaerts valve gear and multiple-bearing crossheads, it still carried Baker gear and an old-style crosshead when photographed in low sunlight on December 13, 1946 at Lordsburg, New Mexico. Regardless of all this upgrading, the B-1s were doomed; they were simply an emergency measure until SP could take delivery on new diesels. In six short years, the 2-8-4 era on the SP came and went.

(A. C. Phelps)

Santa Fe #4193-4199 were selected from the same B&M 2-8-4 fleet as the SP. No immediate changes are apparent on this very early image of Santa Fe #4199 (B&M #4018) taken at Kansas City on October 7, 1945. Like the SP, the Santa Fe had its own "coal territory," but it appears that this particular locomotive was purchased with a much larger tender than those sold to the SP. The B&M had converted it to Walschaerts valve gear. *(Ken Casford)*

At French Camp, where in earlier times the SP crossed the Tidewater Southern passenger tracks south of Stockton, the #3506 gets underway on August 5, 1950 with a train of Pacific Fruit Express refrigerator cars. The head brakeman is still "up top," making sure that this train is ready for the main line. In contrast to the #3504 across the page, #3506 has been converted to oil fuel (with a "whaleback" tender tank), had its booster and feedwater heater removed, and received a corrugated sheet steel pilot. *(Southern Pacific)*

According to SP records, the #3507 was rebuilt from coal to oil—and lost its Coffin feedwater heater—at El Paso on March 1, 1950. Less than three weeks later, gleaming in new paint, #3507 had its moment of stardom on the only known passenger operation of an SP 2-8-4: an excursion in northern California, from Oakland to Tracy and return. The March 1950 fan trip pauses for pictures at Livermore. *(Arthur L. Lloyd)*

Above: Everyday duties in the San Joaquin Valley became the final curtain call for the ten ex-B&M 2-8-4s. The #3508 departs Stockton on August 23, 1950 with the first section of westbound train 423. *(Norman W. Holmes)*

Left: Two weeks into its career as an oil-fired "California Berkshire," #3504 stops over at Indio on February 13, 1950, on its way to a new assignment on the San Joaquin Division. El Paso Shops converted the valve gear from Baker to Walschaerts and gave #3504 a standard SP corrugated pilot, as well as a Pyle headlight. The original tender underframe and trucks were retained in the rebuilding, but the tender body was replaced by a semi-cylindrical whaleback tank from a scrapped articulated locomotive. *(Allan S. Youell)*

Both the #3500 and 3501 operated in California with their Coffin feedwater heaters intact; however, their trailing truck boosters were removed before they left El Paso. The #3501, photographed in December 1949 at Los Angeles, features Walschaerts valve gear and a multiple-bearing crosshead, but has yet to have its original B&M-style headlight replaced. *(G. M. Best)*

On May 23, 1950, the #3501 visited Sacramento Shops to have its feedwater heater removed and an SP-style visored headlight installed. In June 1950, it was back in service and is pictured on the ready track at Fresno. *(Phillips C. Kauke)*

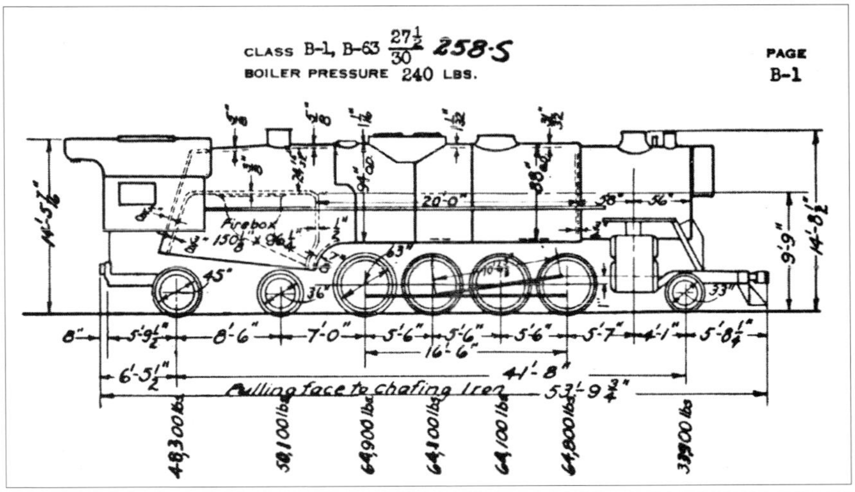

Below: Once the war was over and railfan photographers could be civilians again, one common goal was to capture images of locomotives due for early replacement by diesels. The 3500s were well recorded in their coal-fired days, but few action images survive from their year-and-a-half of California service; by mid-1951, all ten were on the dead line. On September 1, 1950, extra 3504 pauses under a Stockton signal bridge. *(Norman W. Holmes)*

The line between Oakland and Tracy featured helper districts both in Niles Canyon and Altamont Pass, with a brief flatlands respite between Pleasanton and Livermore. On May 10, 1955, extra 3666 east gets underway after taking water at Livermore for the climb ahead over Altamont Pass. The smoke plume in the distance is from cab-ahead helper #4212. *(John E. Shaw, Jr.)*

Chapter Ten
2-10-2: The "Decks"

The 2-10-2, or "Santa Fe Type," was developed on that other railroad, with initial deliveries (the 900 class) arriving in 1903-04. Meanwhile, the Harriman Lines pursued the Mallet concept—but in 1917, SP developed its own 2-10-2, assigning the class letter "F," rather than "SF." Engine crews called them "Decks," for "Decapod" (ten feet in Latin).

Alco built twenty-three class F-1 2-10-2s in 1917-18 for SP (#3600-3610), T&NO (#970-981) and the Arizona Eastern (#1001). The dozen T&NO 2-10-2s were to be the only examples that road received new. SP then returned to Baldwin for all remaining 2-10-2 orders; between March and July 1919, Baldwin built forty-one F-1s, #3611-3651.

The F-1s were built with 63-inch drivers and 27½x32-inch cylinders and weighed 174 tons. However, the #3600 was built as class F-2, with 57-inch drivers and 26x32-inch cylinders. This experiment did not work out and the #3600 was rebuilt in January 1920 as an F-1. Later F-3, F-4 and F-5 locomotives rolled on 63-inch drivers and carried 29½x32-inch cylinders; they weighed about 197 tons. Only the F-4 and F-5 classes carried trailing-truck boosters.

Fifteen F-3 locomotives, #3652-3666, were built in January 1921. Simultaneously, SP decided to transfer Arizona Eastern F-1 #1001 to the Pacific Lines; in order to keep all the F-1 engines in a single number series, new F-3 #3652 was renumbered to #3667 while the AE engine took its old number.

Fifty F-4 2-10-2s, #3668-3717, were built during January through May of 1922. This was a period of business recession when SP locomotive orders outpaced traffic growth, so #3668-3687 were held at Baldwin until May 22, 1922, when they were shipped west as a solid 20-locomotive train, the *Prosperity Special.*

Satisfied with the concept, SP returned for the F-5, signing for fifty-one (#3718-3768) to be built between September 1923 and March 1924. One final 2-10-2, #3769, was delivered in July 1924, becoming SP's rolling test bed for advanced performance concepts. It was built with Uniflow valves and Walschaerts valve gear, then in 1929 converted to Caprotti valve gear and poppet valves. None of these modifications worked well; in 1937 the #3769 was rebuilt as a standard F-5.

Even before its newer 2-10-2s had been delivered, SP disrupted the series by transferring all but one of the class F-1 Alco machines to the Texas & Louisiana Lines. The Pacific Lines number sequence between #3600 and 3769 was never "solid;" in fact, eighty Pacific Lines 2-10-2s migrated to the T&NO between 1923 and 1950. Only a few returned to home rails.

In 1922 and 1923, sixteen F-1s were transferred to the Texas & Louisiana Lines as #982-996. Fourteen more F-1s followed in 1927-29 as #955-961, 965-969 and 998-999. One last F-1, #3619, moved east in 1949, becoming T&NO #954.

No F-3 locomotives were ever sent to the T&L Lines. In 1949 and 1950, four F-4s became T&NO #940-943. The first ten F-5s were transferred to the T&NO in 1929-30, becoming #912-921; six more were transferred east in 1937, as T&NO #906-911. Finally, between 1948 and 1950, twenty-nine F-5s became T&NO #900-905, 922-939 and 946-949.

Still capable of useful work after World War II, the 2-10-2s were displaced as diesels arrived. Banished from bad-water territory, they congregated on the Coast, Western and San Joaquin Divisions to run out their miles. All that remained on the Pacific Lines served through the Korean War years, but were rapidly retired after 1953.

Of the fifty-three original class F-1 2-10-2 locomotives, thirty-three were transferred to the Texas Lines between 1922 and 1929, leaving scattered numbers between #3611 and 3652 to work on the Pacific Lines. Arriving before SP's adoption of the booster engine and the Delta trailing truck, the F-1s suffered in comparison to more modern 2-10-2s. Fitted with a whaleback semi-cylindrical tender, #3627 is serviced at Colton on October 26, 1947. *(Allan S. Youell)*

F-1 #3634 entered service in July 1919 and would remain on the active roster until the end of steam freight operations in November 1956. Carrying one of SP's latter-day corrugated steel pilots, the #3634 is parked on the westbound ready track geographically northwest of Taylor Roundhouse at Los Angeles on May 14, 1947. *(Allan S. Youell)*

Firebox maintenance was simplified by moving the cab back and beveling the front cab wall to allow access to more firebox staybolts from outside the cab, rather than via cramped quarters inside the cab. SP installed so-called "sport cabs" on two class F-1 2-10-2s, #3619 and 3643. Sparkling in fresh paint at the Brooklyn roundhouse on May 29, 1948, #3643 is a 1919 Baldwin product. *(Don H. Roberts)*

A few Alco S-2 switchers delivered during World War II were put to work in Arizona as road helpers, where they could economize on boiler water. They were given train number indicators, road pilots and even ladders and grabirons to facilitate changing train numbers. On February 19, 1946 near Gila Bend, Arizona, #1334 assists westbound 3rd854, powered by 2-10-2 #3692 and its auxiliary water tank. *(Phillips C. Kauke)*

A classic image at San Luis Obispo on March 17, 1952 finds F-4 #3679 entrained ahead of a GS-4 as the helper on westbound *"Sad Sam,"* mail train #71. SP was able to eliminate steam helpers from Cuesta after about 1954, but until 1956 steam road engines were assigned to this train. *(Allan W. Styffe)*

Assigned to the San Joaquin Division, F-3 #3654 waits at Fresno on July 9, 1956. In the period 1937-1945, this class (#3653-3667) was modified with disk main drivers. Alkaline water impurities caused boiler water foaming, leaving an obvious stain on the boiler jacket; daily locomotive baths were just a memory by this late date. *(Herb Cearley)*

By contract, the Arizona-El Paso-Tucumcari main line of the former El Paso & Southwestern was limited to coal fuel, but the EP&SW had no freight locomotives larger than a 2-8-2. Rather than buy new coalburners upon taking over the EP&SW in 1924, SP chose to convert oil-fired 2-10-2s #3653-3667 to coal at El Paso shops in 1925-26. Oil tanks were removed from their Vanderbilt tenders and Elvin stokers installed. Once the 2-8-8-4s arrived in 1939, there was less need for coal-fired 2-10-2s and they were gradually converted back to oil. The #3656, at El Paso on January 12, 1940, will remain a coalburner until 1947. *(Allan S. Youell)*

"Native" Great Basin water in northern Nevada and Utah required expensive treatment before it was suitable for locomotive boilers, so the SP resorted to auxiliary water cars coupled behind locomotive tenders to extend their range between water stops. Once larger tenders became available, the small water cars were used less frequently. Pictured at Sparks on May 29, 1948, F-3 #3663 leads a small tender and a water car. *(Allan S. Youell)*

Shasta Division-assigned road engines were often fitted with temporary snowplow pilots during the winter. F-4 #3702, built by Baldwin in 1922, has been added on February 23, 1952 at Dunsmuir to assist 4-8-4 #4414 on the Portland-bound train #20—by this date little more than a makes-all-stops mail train. In June 1954, the #3702 will be vacated at the Dunsmuir roundhouse and subsequently sold for scrap. *(John C. Illman)*

Class F-4 #3695 is pictured on May 18, 1947 at Los Angeles. The odd tender was considered a class 120-C-1 and was a former coal tender converted from an oil fuel class 120-C-1 car. (The former coal tenders were not assigned to a different class, but they should have been, since they looked very different.) *(Allan S. Youell)*

A peddler freight from the Salinas Valley embarks on the last leg of its journey to San Luis Obispo, departing westbound from Santa Margarita in January 1948. Ahead lies the easier approach to Cuesta, but the downgrade on the other side remains a challenge for this engineer. *(Fred H. Matthews)*

Pacific Lines class F-4 locomotives #3668-3717 arrived in 1922, many of them making their debut on the famous *Prosperity Special* publicity train. F-4 #3715, dating from May 1922, was assigned to the Coast Division in its final years, being vacated at Bayshore at the close of steam operations. In happier times, this machine (fitted with a clamshell exhaust splitter) is pictured at the Bayshore sandhouse. *(Strapac collection)*

After the end of World War II, the Union Pacific quickly retired its fleet of secondhand 2-8-8-2s purchased from the Chesapeake & Ohio, but sold many very large tenders to the SP. Initially, SP put them to work behind 2-10-2s on the Salt Lake Division, to minimize water stops in this traditionally bad-water area. Class 252-R-1 tender #9196 rides behind F-4 #3716 at Carlin on October 2, 1949—and will go on to serve behind #3811 and 3808 after those 2-8-8-4s are converted to oil. *(G. M. Best)*

Baldwin delivered fifty-one booster-equipped class F-5 locomotives to the SP between September 1923 and March 1924, numbered 3718-3768; they were used on heavy freights in territory that didn't justify use of an articulated. Fifteen were transferred to the T&NO in 1929-37, forever denying their images to California photographers. When road freight diesels began arriving on the SP, thirty more moved to Texas in 1949-50, leaving only #3718, 3727, 3734, 3744-45, 3758 and 3765 working in modern times on the Pacific Lines. The #3758 hosts the first section of train #785 at Fresno on January 27, 1952. *(Stan Kistler)*

Class F-5 #3734 emerged from Baldwin in November 1923. Unlike most of this class, it remained a Pacific Lines engine and was never given a T&NO number. Running out its last miles on the San Joaquin Division, the #3734 waits for its next assignment at Fresno on November 30, 1954. It will be vacated in May 1955 at Bakersfield.

SP's West Valley Line provided an alternate route between Tracy and Fresno, but was not used by daytime passenger trains—so vintage photographic images are scarce. In 1954, eastbound second 432 has stopped for water at Los Banos, approximately midway on this obscure line. Class F-5 #3718 is equipped with a clamshell smoke deflector and an uncommon (for this class) smokebox-mounted bell. Nothing in this photo remains today—even the track and the palm tree are gone. This area is now known as the "Los Banos Rail Trail." *(Charles H. Givens)*

Trailing an auxiliary water car at Sparks on May 20, 1948, F-5 #3752 was at this moment assigned to the Salt Lake Division. In April 1950, arrival of new F-unit sets will release #3752 from this duty and allow it to be transferred to the T&NO, where it will be renumbered to #939—disappearing forever from the Pacific Lines roster. As T&NO #939, it will be scrapped in April 1953 at Houston. *(Allan S. Youell)*

Well into the diesel era, SP operated shuttle trains to move logs from railhead to sawmills. Near the California-Oregon border on the Siskiyou Line, extra 3681 east arrives at Hilt in June 1951, delivering logs for the large Fruit Growers Supply mill located here. Many of the logs in this train were loaded on the Yreka Western and handed off to the SP at Montague for the 26-mile uphill ride to Hilt. Once these dozen log flats have been dropped off, the light engine and caboose could either proceed a few miles to Ashland or return to Dunsmuir. *(Charles H. Givens)*

F-1 locomotives active on the Pacific Lines in the 1940-50 era:
3611-3618, 3619*-3620, 3625, 3627, 3629, 3634, 3636, 3638-3639, 3643*, 3647, 2nd 3652.

F-5 locomotives transferred to the T&NO before 1940:
3725, 3731, 3733, 3735-3736, 3738-3741, 3746-3747, 3749-3750, 3754, 3759.

F-5 locomotives transferred to the T&NO 1948-50:
3719-3724, 3726, 3728, 3730, 3732, 3737, 3742-3743, 3748, 3751-3753, 3755-3757, 3760-3764, 3766-3769.
** Modified with sport cabs.*

Southern Pacific F-5 #3738 became T&NO #910 in October 1937, operating there until April 1951, when it was leased back to the Pacific Lines—but for some reason retaining its T&NO number. On March 20, 1953, it was renumbered back into the Pacific Lines series, but not given back its original number. It appears at Los Angeles in June 1953 as #3770, a number that it will carry for its final eighteen months of service. *(G. M. Best)*

A westbound Siskiyou Line extra freight rattles through Creswell, Oregon, on August 1, 1949. Not far from its terminal point at the Eugene Yard, between Cottage Grove and Springfield, the #3762 will have an easy run from here on out. Over-lubrication of a balky bell ringer mechanism has discolored aluminum paint on the smokebox door. In 1950, the #3762 will move to Texas and finish out its years as T&NO #930. *(Fred H. Matthews)*

The only modern-era new articulated locomotives built for the SP with cabs in the normal position were the dozen 2-8-8-4s erected in 1939 for use on the Rio Grande Division, where SP was contractually obligated to use coal for fuel. Once that contract expired, SP was free to convert these big Lima machines to oil fuel, dieselize the Rio Grande Division and transfer the not-yet-amortized 2-8-8-4s to districts that were still using steam. One of those repositioning moves is taking place in the improbable locale of the San Joaquin Valley—at Delhi, south of Modesto—on April 12, 1953. Lima #3809 is "just passing through," powering the second section of westbound freight #423, on its way to a new home on the Modoc Line. The #3809 would never again visit this location as a live locomotive. *(Guy L. Dunscomb: Menke collection)*

Chapter Eleven
Mallet and Articulated

The term "Mallet" denotes an articulated locomotive (one with a vertical hinge in its frame) that uses steam twice—first in high-pressure cylinders serving the driving wheels nearest the firebox and cab, then second in cylinders operating the drivers near the smokebox. The steam loses considerable energy between the primary and secondary cylinders, requiring a larger cylinder volume in the low-pressure cylinders to deliver work. An "articulated" locomotive uses steam only once; each pair of cylinders is supplied with "new" steam. SP converted its early Mallets to simple articulateds during the 1920s and 1930s.

Southern Pacific's latter-day Mallets and articulated locomotives (both vintage and contempory) are discussed in this Chapter.

In 1909, the Harriman Associated Lines developed specifications for a 2-8-8-2 freight locomotive as a replacement for multiple 2-8-0s on freight trains in mountain districts. Initial deliveries were SP #4000-4001 (Baldwin April 1909), UP #2000-2002 and O-WR&N 450-452 (Baldwin November 1909). The builder's performance claims proved valid, but smoke and heat in tunnels and snowsheds created an intolerable problem for engine crews. Backing the Mallet through tender-first solved the gas problem, but now the engineer was on the wrong side for signals—and forward visibility was compromised.

SP's first fifteen production model 2-8-8-2s were on order for delivery from Baldwin in November 1909, but SP intercepted the original order and sent a change order to turn the locomotive around in relation to its tender. Mallet Compounds #4002-4016 and subsequent #4017-4048 were all built with cabs out front. Soon after delivery, SP pressurized the tenders to push fuel "uphill" to the firebox at the front of the locomotive.

Passenger service between Roseville and Sparks was to be similarly upgraded with Mallet power. In 1911, SP ordered a dozen 2-6-6-2 compound cab-aheads as class MM-2 (Mallet Mogul), numbered 4200-4211. They did indeed reduce the need for doubleheading passenger trains on The Hill, but, after a series of spectacular derailments, SP decided that they were unstable as built. To provide better tracking, they were almost immediately rebuilt with four-wheel leading trucks. In 1938 they were renumbered #3900-3911.

Both the 2-8-8-2s and 4-6-6-2s shared design details typical of the Common Standard era, but by the 1920s, modern locomotive technology had passed them by. They were slow, difficult to maintain and relatively wasteful of fuel. In an effort to increase their productivity and extend their service lives, between 1928 and 1936 SP rebuilt all the 4-6-6-2s and all but three of the 2-8-8-2s as simple locomotives. After little use during the Depression years, these old machines were worn out in continuous service during World War II, becoming the first casualties among larger steam locomotives, as diesels began arriving. The 3900s were all scrapped 1946-48, while the 2-8-8-2s were retired in the same time frame. After a flurry of post-war photography, they were all gone by 1949.

Seeking a modern, high-performance steam locomotive to operate on the Rio Grande Division where coal—not oil—was used as fuel, SP reverse-engineered a locomotive similar in specifications and performance to its successful 4-8-8-2 cab-ahead simple articulateds, but fired with coal and with its cab and firebox in the conventional location adjacent to the tender. Bids were solicited and Lima won the contract, erecting twelve 2-8-8-4s, #3800-3811, in late 1939. Class AC-9 featured vestibule cabs and skyline casings. They were converted to oil fuel at El Paso between March and August 1950; all except the #3800 were transferred in 1953 to the Modoc Line in California. They were scrapped 1954-56.

Two Alco compound 2-6-6-2s were purchased in 1943 during the war emergency. They came from the Verde Tunnel & Smelter Railway in Arizona and were promptly renumbered SP #3930-3931. Their tiny 9,000-gallon tenders were soon rebuilt to 12,000 gallon capacity. Exclusively Los Angeles Division engines, they served about a decade on Harbor Drags and in helper service before retirement in 1951 and 1954. They were the last compounds to work for the SP.

Among the most graceful of all modern articulated locomotives, SP's dozen AC-9s were built to burn New Mexico coal. Lima delivered #3800-3811 in October-November 1939, introducing skyline casings and vestibule cabs for SP freight locomotives. Probably no parts were interchangeable with cab-ahead locomotives, but most important dimensions were repeated (the 3800s' firebox was longer). Until the Rio Grande Division went to oil fuel for steam locomotives in 1949-50—and to diesel power in 1953—these elegant machines were generally confined to Tucumcari-El Paso operations. *(Lima Locomotive Works)*

In 1953, eleven of the twelve AC-9s were reassigned to the Modoc Line, crossing northern California, Oregon and Nevada. Their primary backshop was Sparks, Nevada, but they worked out of Alturas. There were no tunnels or snowsheds or other restrictive clearances on this line (opened in 1929), but SP took the extra step of stenciling a notice on the front cylinder casing that read "This Engine Excess Width." The last two 2-8-8-4s to operate, #3804-3805, became the subject of intense photographic attention in their last summer; the #3804 is pictured on July 30, 1955 at Alturas. It would be vacated the following April. *(Stan Kistler)*

It's January 3, 1953 at El Paso; in another few months, oil-fired #3803 will depart the rapidly dieselizing Rio Grande Division and take up new headquarters at Sparks, Nevada. Alco built that huge class 252-R-1 tender for a Chesapeake & Ohio 2-8-8-2, which was sold to the UP during World War II. SP bought fourteen large tenders when the UP was retiring its ex-C&O 2-8-8-2s after the war, assigning them initially to 2-10-2s, but eventually using them to replace the original coal tenders running with its AC-9s. Less than fourteen years old, this Lima 2-8-8-4 will put in a few more months of work—and generate some thousands of ton-miles—but will be vacated from the roster (at Sparks) before the end of 1954. *(W. C. Whittaker)*

Coal is piled high in the 3800's tender at El Paso on October 19, 1947. Espee operated the 3800s as coal-burners for eleven years; in 1950, when they were converted to oil fuel, their original Buckeye-truck tenders were either set aside for use as water cars or scrapped. Replacement tenders purchased third-hand from the UP boasted even greater capacity. In the early fifties, the #3800 would be set aside as a parts source and would never be transferred to California; its was vacated on June 12, 1953 and scrapped July 29 at El Paso.

(Leo Munson collection)

Rio Grande Division train #992 is pictured at El Paso on September 4, 1952. Lima 2-8-8-4 #3802 was converted from coal fuel to oil some two years prior to this picture, but full dieselization on the division (and resulting transfer of the 2-8-8-4 fleet to California) is still some months in the future. The eight tank cars in line immediately behind the tender appear to be lettered for the Army Transportation Corps, perhaps delivering aviation fuel to the many airbases in West Texas and southern New Mexico.

(W. C. Whittaker: Menke collection)

When the Rio Grande Division was converted from coal fuel to oil in 1949-50, the ten-year-old conventional cab Lima 2-8-8-4s were rebuilt at El Paso General Shops. Between March and May of 1950, they were converted to oil fuel, receiving third-hand Chesapeake & Ohio tenders that had themselves been rebuilt for oil fuel. The AC-9s were finally displaced from their home rails in New Mexico in 1953 and transferred to the Modoc Line in northern California, a territory with neither tunnels nor snowsheds, to run out their final miles. Railfans organized a fan trip in May 1954, utilizing the #3809 (pictured at Wendel) on the only recorded passenger operation of this class. *(W. C. Whittaker)*

Purchased from Baldwin in 1911 with high hopes for mountain passenger service, compound 2-6-6-2s #4200-4211 were soon discovered to be unstable and prone to derailment; they were quickly converted to 4-6-6-2 with a newly-engineered front truck. They were rebuilt to simple operation between 1928 and 1937 and were renumbered in March 1938 to #3900-3911 to make room for planned deliveries of 4-8-8-2s. They spent their last years on the Portland Division, mainly shuttling freight between Brooklyn and Eugene. Not long after this image was snapped at Brooklyn in 1947, #3901 would be cut up for scrap. *(Don H. Roberts)*

As soon as SP began receiving Electro-Motive F3 units in 1947, management began sweeping old freight locomotives from the roster. The 3900-series 4-6-6-2s and the 4000-series 2-8-8-2s would all disappear before 1950. Only ten months away from final retirement, #3908 waits at Brooklyn on May 27, 1947 while the fireman puts up the indicators for Eugene-bound train #665. That gray box below the front number plate contains a block of ice for drinking water. *(Richard H. Kindig)*

In August 1935, a teenage boy with a camera was rewarded by the engine crew of extra 4026 west as the train clomped through the Eugene passenger station. Many Portland Division-assigned articulateds were equipped with small pilot plows like this one. While 4-6-6-2s in the 3900 series could easily shoulder the load in the Willamette Valley, it took at least sixteen drivers (and probably a helper) to move freight beyond Eugene on either the Siskiyou or Cascade Lines. *(E. M. Leo: Menke collection)*

The first section of Eugene-bound train 663 blasts through Salem, Oregon, on July 21, 1947. Conveniently for the photographer, the fireman has chosen this location to sand out the flues. AM-2 #3902 has been fitted with a modern pilot and sports new paint, but is running on borrowed time. *(W. C. Whittaker: Menke collection)*

No details are known about the setting or date of this picture, but it would appear to be in an urban area in Oregon about 1938 to 1940. After World War II, the #3907 will be modernized with a solid pilot and a Pyle headlight. *(W. C. Whittaker collection)*

The Portland Division had an oversupply of old cab-ahead 4-6-6-2 and 2-8-8-2 locomotives in 1940, but once war broke out, every one of them was hastily shopped and returned to service as soon as possible. Only urgent maintenance was performed during the war years. Class AM-2 4-6-6-2 #3905, built in July 1911 as compound 2-6-6-2 #4205, became a simple engine in January 1929 and was renumbered in March 1938. By July 1942, it had the obligatory headlight dispersion shield and was leading an extra through Albany, Oregon. Five years later, it would be scrapped. *(W. C. Whittaker: Menke collection)*

After the war was over, the 4-6-6-2s remained on the "valley" portion of the Portland Division, powering shuttle trains between Brooklyn Yard (near Portland) and Eugene. Pictured at Salem on July 21, 1947, the #3909 has received new paint and new lettering, as well as a Pyle headlight and a modern corrugated steel pilot. It has less than six months until a date with the scrapper at Brooklyn. *(W. C. Whittaker)*

As larger locomotives were scrapped, more 16,000-gallon Vanderbilt tenders became available, allowing SP to set aside the MM-3s' rectangular tanks, which even after rebuilding from 9,000 gallons were still only good for 12,000. By 1952-53, these old-technology Mallet compounds were kept "close to home" on the Los Angeles Division, working heavy locals to Colton and Wilmington and occasional helper chores on Beaumont Hill.

The two former Verde Tunnel & Smelter compounds, purchased in February 1943, were confined to the Los Angeles Division for their entire decade-long SP careers. While useful on drag freights to Colton or the Harbor or for helper chores east from Colton, they were too slow for mainline assignments. Secondhand 2-6-6-2 #3930 often posed for railfan photographs on this track at Taylor roundhouse; this image (by an unknown photographer) was snapped on April 27, 1947. *(SPH&TS collection)*

A long-standing tradition in both northern and southern California saw railfan organizations requesting odd locomotives for excursion trains. Alco 2-6-6-2 compounds were relatively common on the Western Pacific, but there were only two on the SP (and they served only about a decade); hence, on April 16, 1950, the freshly painted (but slow) #3930 hosted a fan trip eastward from Los Angeles. Almost ready for the return leg of the trip, #3930 and a short passenger consist wait at Beaumont. *(F. J. Peterson: Miller collection)*

The awkward lines of its original tender (rebuilt by SP from 9,000 gallons to 12,000) emphasized by obsolete "Southern Pacific Lines" lettering, 1920-vintage Alco compound #3931 awaits its next assignment on January 4, 1947 at Taylor Yard with the Pacific Fruit Express shop building in the background. *(Leo J. Munson collection)*

Once traffic receded to more normal levels after World War II, SP had the luxury of redeploying older, less powerful locomotives to less demanding services. The 4000-series 2-8-8-2s, Baldwin-built as compounds between 1909 and 1913, were rebuilt to simple in the 1920s and 1930s, extending their useful careers through the war years—but after the war, they were banished to Oregon and the Modoc, where they ran out their last miles in relative obscurity. Photographer Richter has focused the fireman's full attention in this perfectly lighted image at Sparks, Nevada, on June 2, 1947. *(D. S. Richter)*

The more lubricated moving parts that were detached, the easier it was to switch a dead locomotive to the scrap track—which is where this 1912 Baldwin is going. Main rods, piston rods and valve gear links have all disappeared from otherwise intact 2-8-8-2 #4032 at Brooklyn in this undated photo. SP records state that #4032 was vacated in August 1949, one of the last of its class. *(Don H. Roberts)*

By 1940, most of the old 2-8-8-2 cab-ahead locomotives had been reassigned away from the Los Angeles Division, but the 4009, 4016 and 4024 remained in southern California. They were useful as helpers and on drag freights, including trips to Wilmington (Los Angeles Harbor). AC-1 #4009, built by Baldwin in 1909 and rebuilt to simple in May 1929, poses eleven years later at the Harbor Belt Line yard in Wilmington. It would be scrapped in 1948. *(Howard L. Fairchild)*

Transferred north from the Los Angeles Division during World War II, class AC-2 #4024 joined many of its brethren on the Modoc line. It is pictured in postwar paint (in the company of a very useful auxiliary water car) on June 1, 1947 at Canby, west of Alturas in far northeastern California. The #4024 was built in May 1911 as a Mallet compound and converted to simple operation in August 1928. It will be vacated in another fourteen months and scrapped in February 1949 at Sacramento.

(W. C. Whittaker)

The engine facilities at Alturas on the Modoc Line became the post-World War II refuge for SP's old 2-8-8-2 cab-ahead locomotives, and thus a natural magnet for Bay Area railfan photographers. On the same June 1, 1947 that Wil Whittaker photographed the #4024 on the previous page, Douglas Richter was capturing this image of the #4020 at Alturas. Latter-day modifications include a Pyle headlight and a corrugated steel pilot.

As an experiment in March 1928, SP rebuilt compound 2-8-8-2 #4028 to simple operation, discovering new life in old Baldwins—in fact, a simple, superheated 2-8-8-2 was just as productive as the vaunted three-cylinder 4-10-2s and easier on the rails. Their low drivers held them to slower speeds, but they more than earned back their cost of rebuilding. In 1947, dusty #4042 waits at the Eugene passenger station for its next assignment, perhaps helper duties on the Cascade Line. Large tender lettering, a splash of aluminum paint on the front panel and a replacement Pyle headlight will be #4042's final modernizations; it will be vacated in March 1949.

Of the four dozen 2-8-8-2s (#4000-4048), only three (#4011, 4013 and 4022) were never rebuilt to simple and were scrapped before World War II. The remaining 45 were worn out during the War and scrapped between 1946 and 1949. Prior to World War II, Portland Division-assigned 2-8-8-2s were typically fitted with SP's familiar plow pilot and carried large headlights. In May 1936, the second section of westbound train 661 is about ready to leave Oakridge, Oregon, where the real climb to Cascade Summit begins.
(Herbert L. Arey: Menke collection)

A collision, perhaps as early as 1941, destroyed the #4045's original cab. Rather than rebuild it to original specifications, it was given an approximation of the AC-7 beveled-front cab (and a pilot plow) and sent back to work. Unlike the 4-8-8-2s, however, this rebuild did not include a nose-mounted air horn. The #4045 is pictured in April 1945 at the helper terminal in Oakridge, Oregon. It would be one of the first 2-8-8-2s to be scrapped after the war. *(H. G. Burnett)*

The everlasting hills and monumental red rocks are still there at Ravenna in the 21st century, but steam-powered SP freights no longer stop at this Soledad Canyon location to refill tenders with water. In 1947, westbound 2nd 811 gets underway on the 2.24% grade on its way to the summit at Vincent, nine miles further uphill. New F3 diesel locomotives are on order, but AC-5 #4116 has yet to receive aluminum trim paint or a solid corrugated steel pilot. *(Thomas Hotchkiss)*

Chapter Twelve
Articulated Consolidation (AC): 4-8-8-2

Southern Pacific's signature freight locomotive, the cab-ahead 4-8-8-2, was the result of a technological breakthrough in locomotive theory—the result of designers realizing that locomotive productivity could be better enhanced by enlarging firebox and boiler capacity (that is, steam generating capacity), rather than by increasing the number of cylinders or their dimensions (the steam-using capacity). This was the basic premise of Lima's "Super Power" concept—that boiler horsepower was more important than cylinder tractive effort.

In the mid-1920s, SP bought 4-10-2 locomotives expecting that they would supplant compound 2-8-8-2s between Roseville and Sparks. However, the new machines were not a success, so SP planning turned back to articulated locomotives. Conventional wisdom implied that the boilers of those 2-8-8-2s did not have the steam-generating capacity to power four simple cylinders, so in June 1927 SP decided to test this assertion by rebuilding a compound 2-8-8-2 (#4041) with simple cylinders. While generating somewhat less tractive effort, the simple #4041 was faster over The Hill than either a 4-10-2 or a compound 2-8-8-2.

Once the merits of simple articulation were recognized, SP rebuilt the rest of the 2-8-8-2s and designed a new locomotive, class AC-4, that would use a still larger firebox supported by a four-wheel leading truck. The first ten 4-8-8-2s (#4100-4109) were delivered in late 1928 and early 1929 and were an immediate success. Re-orders followed for sixteen in 1929 (#4110-4125) and twenty-six more (#4126-4150) in 1930. During the Depression, no new locomotives were required, but as soon as business picked up in late 1936, SP ordered 25 class AC-7 locomotives (#4151-4176)—very similar in specifications to their predecessors but boasting beveled cab fronts and large, 22,000-gallon rectangular tenders.

The international situation in 1938-39 led SP management to anticipate another World War; as a result, in 1939, class AC-8 (#4177-4204) was delivered. War did indeed come and the government quickly clamped down on locomotive deliveries—especially new diesels. SP was satisfied with its proven cab-ahead steam freight locomotive and was able to obtain ninety (#4205-4294) from Baldwin between 1942 and 1944, avoiding waiting in line for EMD FT diesels. There were few design enhancements on these war-era 4-8-8-2s; SP was fortunate to obtain as many as they did. While there were larger and more powerful locomotives built in this same time period, SP and Baldwin got this design right.

While they weren't allowed on most branches and they could not be double-headed on heavy freight trains due to coupler strength questions, the 4-8-8-2s were an extremely successful design. They could haul heavy passenger trains (especially on Tehachapi) and broke the freight bottlenecks on both the Sierra and Beaumont.

Classes AC-4 through AC-12 worked side-by-side with Electro-Motive F3 and F7 locomotives from 1947 until November 1956, when deliveries of new GP9 locomotives finally allowed SP to discontinue steam freight operations. The first modern AC to be retired was the #4110, vacated in December 1952. The last to operate was class AC-11 #4274, which powered a farewell excursion from Sacramento to Reno (Sparks) and return on November 30 and December 1, 1957. Between 1956 and 1958, all but one were sold for scrap.

The #4294, the newest cab-ahead and SP's last new steam freight locomotive, was donated in October 1958 to the city of Sacramento for permanent display. In May 1981 it was moved into the new California State Railroad Museum and resides there to this day.

Southern Pacific's first modern, simple articulated locomotives were the ten class AC-4 engines, #4100-4109, built August through October 1928. Not yet ready for 250 pounds of boiler pressure, the railroad specified 235 for this and the following order; otherwise, the essential pattern set in 1928 would be followed through the 1944 orders. The #4100 entered service on October 27, 1928 and is pictured at Los Angeles almost 19 years later, on October 8, 1947. *(Leo Munson collection)*

On the eve of World War II, freight traffic was booming in southern California; a favored neighborhood for action photography was SP's busy double track between Los Angeles and Burbank Junction. For instance, the first section of westbound train 813 for Bakersfield has barely hit its stride between Los Angeles and Glendale after departing Taylor Yard on October 15, 1941. AC-5 #4115 models the original style pilot. *(Phillips C. Kauke)*

Faced with the reality of a sixteen-foot-long firebox, SP opted for firebox insulation and jacketing all the way down the side, retaining this feature on every cab-ahead. Southern District-assigned ACs typically did without exhaust splitters, since they operated on lines with no snowsheds. The #4109 is pictured at Taylor Yard in Los Angeles on May 7, 1940; the tender still carries "Southern Pacific Lines" lettering, but in the future, SP will simply omit tender lettering on hard-working cab-aheads. *(Allan S. Youell)*

One innovation used on the later AC engines was access to the train number indicator boxes from inside their larger cabs. Some earlier cabs were modified to feature this convenience, resulting in a slightly different front aspect—less costly than fabricating and installing an entirely new cab. Extra 4107 east works through Martinez on July 11, 1954. *(J. C. Hammond: Shaw collection)*

Stopping for a crew change on March 27, 1941 at Mojave, the third section of Los Angeles-bound train 812 has AC-5 #4111 for power. Profiled atop the tender is a pressure lid for the oil bunker, which had to be pressurized to force fuel oil all the way up to the cab end of the firebox, where the oil burner was located. At the very rear top profile of the smokebox is a small electric light on a stalk, used at night by the fireman to assess the condition of his engine's fire. *(Allan S. Youell)*

Extra 4122 west works through College Park, between San Jose and Santa Clara, on the afternoon of June 22, 1950. AC locomotives were allowed to run up the Peninsula, but normal practice was to power heavy freights bound for South San Francisco or Redwood City with 2-10-2s, dispatching even heavier trains to Oakland or Tracy using 4-8-8-2s. *(Norman W. Holmes)*

Westbound on train 785 at Mojave on August 17, 1941, AC-5 #4120 shows signs of hard use as the SP ramped-up its operations in anticipation of a full war footing, just a few months in the future. Almost exactly eleven years into its service career since going to work on August 2, 1930, the #4120 still features the early standard of a water cooler outside the front wall of the cab.
(Allan S. Youell)

The last of the old-style 4-8-8-2s, AC-6 #4150, went to work on January 26, 1931. It was rebuilt after World War II with larger cab windows and—more importantly—cast steel frames, eventually receiving a corrugated steel pilot as well. In common with most of its flat-faced articulated brethren, the #4150 was retired in 1955. It was assigned to an eastbound work extra in 1950 near Beaumont.
(Thomas Hotchkiss)

In July 1942, AC-6 #4129 was rebuilt with this close approximation of an AC-7 cab as well as a new corrugated steel pilot. The air horn commonly seen on later classes was not installed. By 1946, it was assigned to helper duties on the Cascade Line, headquartered at the remote outpost of Oakridge, Oregon. The #4129 would serve for eight more years, finally vacated in May 1954.
(H. G. Burnett)

After a depression-era hiatus in 4-8-8-2 deliveries, SP returned to Baldwin in late 1936 with an order for twenty-six class AC-7 4-8-8-2s, featuring the 250-pounds boiler pressure setting introduced on the AC-6 engines, as well as an enlarged cab with a beveled front face. SP designed a new rectangular tender rated at 22,030 gallons of water and 6,400 gallons of fuel oil. As was typical practice with all cab-ahead tenders, the oil tank was pressurized to force oil "uphill" to the firebox. The #4151-4176 were delivered between January and August of 1937; #4174 takes a break in August 1950 at Crescent Lake, Oregon.

Fading light on June 13, 1954 illuminates Los Angeles-bound train 60, the *West Coast,* at the Antelope Valley station sign of Oban, milepost 400 from San Francisco. While a smaller locomotive would suffice at this particular location (where the AC could roll along at an easy—but fuel gulping—fifty miles per hour), only forty miles of Valley floor separated the grades of Soledad and Tehachapi, each of which would challenge AC-7 #4176. *(Ray Ballash)*

In order to maintain a constant 1.8% westbound grade when building the Cascade Line in the 1920s, SP had to engineer a huge S-curve and drill seventeen tunnels on Willamette Pass. With ten miles to go (and almost 800 feet more altitude to gain), extra 4164 west emerges from a tunnel near Cruzatte, Oregon, in December 1950. In this environment, the need for a pilot snowplow is apparent.

(Bert H. Ward)

Latter-day SP steam locomotive maintenance planning focused on "summer rush and winter storage," allowing considerable shop work to be completed in the off-season when there weren't as many perishable extras to dispatch. As a result, even toward the end of their service careers, SP's first-line steam locomotives were kept reasonably clean, if not shiny. Gleaming in new paint at Sparks on July 4, 1955, AC-7 #4168 would work until the end of steam operations sixteen months after this picture.

(Fred A. Stindt)

The remote Modoc Line was originally planned to provide a shortcut for Oregon forest products moving toward eastern markets, bypassing the ride from Klamath Falls down to Roseville and thence over the Sierra to Ogden. Completed in the 1920s, the Modoc never lived up to its economic potential—but in the 1950s provided an opportunity for SP to run off the miles on large articulated locomotives and postpone dieselization a little longer. On August 13, 1954, AC-8 #4191 and AC-12 #4292 have stopped briefly at Ravendale, between Wendel and Alturas on the Salt Lake Division. *(John E. Shaw, Jr.)*

The large 4-8-8-2s were not common on the San Francisco Peninsula, but those ACs assigned to the Coast Division were required to visit Bayshore Shops (near Brisbane) for their mandated maintenance. Not yet assigned to a train on November 24, 1951, AC-8 #4185 (built by Baldwin in late 1939) has steam up and will soon return to active work. *(W. C. Whittaker)*

Beaumont Hill in southern California was one of many helper districts on the SP where trains traveling in either direction required assistance. The summit, in the town of Beaumont, was arranged to turn back helpers either to Colton or to Indio, providing a wye and ample supplies of boiler water. A summer afternoon in 1952 finds AC-8 #4201 in company with a 4-10-2 and another AC in the far distance, perhaps already returning to Indio. *(W. C. Hendrick)*

On July 2, 1955, eastbound extras powered by F7 #6287 and AC-8 #4188 wait to be released from the small Gerber Yard while a railfan special powered by GS-6 #4463 holds the main to take on water and fuel. Built in August 1939, the #4188 would not operate after this summer's rush and will be vacated on March 5, 1956 at Sacramento. *(Norman W. Holmes)*

AC-10 #4232 contributes to the April morning mist and fog in the spring of 1954 in Niles Canyon, just west of Brightside. Eastbound train 408 must still negotiate Altamont Pass, a few miles ahead, before dropping down into Tracy in the San Joaquin Valley. The photographer recalls cutting all of his morning college classes on this occasion to make a little history of his own. *(Charles Givens)*

Fresh from its morning steam bath, AC-10 #4240 waits in the brilliant morning sunlight at Bayshore on May 25, 1956. Through trains on the Coast Division were already dieselized, with new GP9s arriving daily—but SP was hedging its bets by continuing a high level of maintenance for its front-line steam power, anticipating future traffic peaks that fell to diesels to serve. By late summer of 1956, the #4240 would join the long dead line at Roseville, never to run again. *(D. S. Richter)*

Tracy-bound at Niles Tower on May 10, 1955, extra 3666 west benefits from a shove provided by AC-10 #4212. Even a steel caboose was not immune from "accordioning" if entrained ahead of a powerful 4-8-8-2, so SP policy was to cut in the helper ahead of the caboose, even if this required more time and effort to accomplish. *(John E. Shaw, Jr.)*

During World War II, new locomotives (especially diesels) were difficult to obtain. Both the Santa Fe and the Western Pacific were allowed by the War Production Board to buy EMD FT diesels, but the SP chose not to get into that line, instead buying ninety AC-10, AC-11 and AC-12 locomotives (#4205-4294) between 1942 and 1944. The #4220, an AC-10 dating from April 1942, receives a steam bath at the West Oakland roundhouse on January 1, 1956. By late November, the #4220 will be out of work. *(Stan Kistler)*

The night Valley passenger train, the *Owl,* carried train numbers 25-26 until 1946, after which timetable train numbers were changed to 57-58. Always a heavy consist, the westbound *Owl* is pictured in 1945 during its station stop at Glendale, powered by AC-11 #4246. The war-time light dispersion shield is gone, but #4246 has yet to receive aluminum paint to brighten up its cab front. The empty tracks on the far right belong to the Pacific Electric, which interchanged local freight at Glendale with the SP. *(Ralph Lettelier: Miller collection)*

Photographers seeking action images of articulated steam on the SP in 1955 were well advised to make the trek to remote Alturas, on the Modoc Line in far northeastern California. AC locomotives (both 2-8-8-4 and 4-8-8-2) with flue time remaining were transferred to this outpost while new diesels took over on other divisions. It was business as usual for steam railroading at Alturas on July 30, 1955, as AC-11 #4250 backs up to its train. *(Stan Kistler)*

Normally, the ACs operated with a clean fire, but on occasion they were caught polluting, either due to deferred maintenance or the fireman sanding the flues. Heading for Roseville from Gerber on a warm day in July 1953, extra 4269 west is near Lincoln on the East Valley line. *(Charles Givens)*

This photograph is one of hundreds taken on November 30 and December 1, 1957, the last time a cab-ahead operated on the SP. A fan trip from Sacramento to Sparks and back was Southern Pacific's final salute to its trademark freight locomotive; by this late date—a year after the end of regular steam freight operations—only the #4274 remained qualified for service. An unknown photographer joined the throng at Penryn on November 30, 1957 to record this parting image.

The last days of 1951 and early January 1952 delivered the "Perfect Storm" to the Sierra, ultimately bringing SP's Roseville-Sparks operations to a halt and stranding the *City of San Francisco.* Even this definitive steam-powered snow-fighting "train" was bested by snow falling faster than it could be plowed out—and their boilers kept supplied with fuel and water. Sandwiched between two rotary snowplows, the #4290 and another 4-8-8-2 leave Roseville on December 30, 1951, for another assault on The Hill.
(Alan Aske)

By early 1956, cab-aheads would be gone forever from Taylor Yard, as big steam withdrew to the Modoc, Coast, Western and San Joaquin Divisions. In anticipation, photographers paid more visits to the Taylor Roundhouse to capture scenes such as this, as AC-12 #4281 is eased off the turntable on April 9, 1955. *(John E. Shaw, Jr.)*

In their final two years of service, the cab-aheads seemed to sparkle with fresh paint and frequent steam baths—they were going to go out in style. At Watsonville Junction on July 3, 1955, AC-12 #4275 has just arrived with a train of empty Pacific Fruit Express ice refrigerator cars slated for the next load of Pajaro Valley vegetables. The first AC-12, #4275 went to work on October 27, 1943.
(Ray Whitaker)

SP enjoyed considerable operating flexibility on the busy section of track between Burbank Junction and Los Angeles, with double track the entire distance and the additional benefit of yard limits at strategic points, allowing restricted movements without timetable authority. For instance, on days when the inbound *Owl* required two cab-aheads to maintain its schedule, the helper could be cut off at Glendale and run ahead directly to tie up at Taylor Yard, while the train continued on to Union Station. On November 16, 1950, in the first picture, helper #4247 has been uncoupled and indicators have been taken down—it is moving as a light engine within yard limits. Following, the #4203 accelerates train 58 the last few miles into Los Angeles.
(Alan Miller collection)

The second section of westbound freight train 671 will soon highball out of Eugene on August 20, 1950, but in the meantime there are a few leisurely moments for a "brotherhood meeting." The late-model cab on AC-12 #4286 contrasts with the modernized flat-front cab of AC-6 #4134: the water cooler has been removed, but its shelf remains in place, while the number plate has migrated upward to a new location between the cab windows. A corrugated sheet-steel pilot completes the upgrades. *(Dick Wolf)*

The *Owl* was by no means a "luxury flyer," but its saving grace was scheduled appearances in both directions at Glendale during daylight hours—as well as the SP's propensity to load up the train with mail and express. This was one of the few SP passenger trains that could be expected to operate with doubleheaded 4-8-8-2s—a practice SP otherwise avoided. On the afternoon of September 6, 1951, #4289 is doubleheaded with 4257 on Oakland-bound train 57 at Glendale. *(Allan S. Youell)*

While overshadowed by the more scenic Tehachapi Pass to the north, SP's crossing of the San Gabriel Mountains via Soledad Canyon was at least as demanding. The summit of the grade from San Fernando was at Vincent (today a Metrolink park-and-ride stop), a 900-foot vertical climb from Lancaster in the Antelope Valley, some fifteen miles away. On a 1950 occasion, extra 4285 west stops for orders at Vincent before diving down into Soledad Canyon. *(Thomas Hotchkiss)*

A favorite with photographers, the Los Angeles-bound *Owl* was generally heavy with head-end cars and often double-headed. It frequently ran late over Tehachapi; even with Daylight Savings Time, the Owl was scheduled through the mountains before there was much morning sunlight—if it was on time. Running more than a little late on June 12, 1950, train #58 is powered through the triple track at Marcel by Mt-1 #4312 and cab-ahead #4257. *(Phillips C. Kauke)*

Chapter Thirteen
4-8-2: Mountain

In 1923, Southern Pacific took delivery of ten new class Mt-1 4-8-2 locomotives (#4300-4309) from Alco. In contrast to SP's recent Baldwin 4-6-2s weighing 153 tons (91 tons on the drivers) that generated 43,660 pounds of tractive effort, the new 4-8-2s weighed 184 tons (123 tons on the drivers) and generated 57,510 pounds of tractive effort. The new machines averaged 61,500 pounds per driving axle vs. 54,930 pounds per driving axle of the 4-6-2s, but spread that weight over a longer wheelbase.

The Mt-1 engines featured 73-inch drivers and 28-by-30-inch cylinders and relatively low (210-pound) boiler pressure. SP signed with Alco for eighteen more (#4310-4327) for delivery in March 1924.

With the availability of commercially cast, single-piece locomotive frames, the process of building complete new locomotives in railroad shops was considerably simplified. The SP could purchase "engine bed" frames for locomotives to be erected in its own shops and equal the dimensional accuracy and structural integrity of the commercial product. A comparative cost study showed that Sacramento Shops could build new 4-8-2s that would compete on both price and performance with the Alco product. As a result, eighteen class Mt-3 locomotives, #4328-4345, were ordered. The shop's production rate allowed adding up to four new 4-8-2s per month.

In November 1924, after the Mt-1 locomotives had been delivered, Southern Pacific took over the El Paso & Southwestern Railway's half-dozen coal-fired 4-8-2s, briefly EP&SW #410-415. These were also built by Alco, but were somewhat heavier and carried larger cylinders than the SP design. They were given class Mt-2 and numbered 4385-4390. In later years, they were converted to oil.

The class Mt-4, #4346-4366, substituted the Franklin type C-2 tailing truck booster for the earlier type C-1. Cabs were relocated a foot rearward in relation to the boiler, allowing external access to all firebox staybolts.

Between November 1929 and March 1930, Sacramento Shops erected the final ten 4-8-2s, class Mt-5. They were numbered 4367-4376 and followed the same general blueprint, except for the air pumps being moved to the engineer's side of the boiler.

Many 4-8-2s were fitted with SP's distinctive "clamshell" exhaust deflector that reduced the vertical blast against showshed and tunnel roofs. Another common modification was installation of a solid plow pilot, similar to the design used on many SP diesels decades later. SP's 4-8-2s were built with "alligator" type crosshead guides, but many were later fitted with the Laird multiple-bearing guide. Others were modified with disc main drivers that could be counterbalanced more accurately than spoked drivers. Most of them, beginning in 1939, were fitted with sheet-steel "skyline casings," similar to those on the new Lima 4-8-4s, developed to help direct exhaust smoke atop the boiler, away from the locomotive cab. This feature was simply a steel box attached to the top of the boiler.

In 1946, in the days when SP made an effort to assign

(Continued on page 186)

SP did not get around to installing the skyline casing on every 4-8-2 in a timely manner; it took years before the fleet was fully equipped. The #4310, for instance, finally received this cosmetic treatment at Sacramento on May 19, 1950—many months after this October 1948 portrait at Roseville. Performance modifications are in place, notably a 16,000-gallon tender. Many SP 4-8-2s were refitted with disk main drivers and multiple-bearing crossheads, but neither feature is apparent here. This engine will be vacated from the roster in May 1954 and sold for scrap the following July.

By the late forties and early fifties, SP had more than enough 4-8-4s and diesels to power its first-line passenger trains—and could "demote" modern 4-6-2 and 4-8-2 locomotives to divisional freight train service when there were no dedicated freight locomotives available. Even the demands of Commute service on the Peninsula couldn't keep every engine busy; for example, Alco Mt-1 #4309 leads westbound train #429 toward Santa Clara near Sunol and Niles Canyon on October 5, 1952. *(John C. Illman)*

SP took delivery of its first ten 4-8-2s, #4300-4309, in late 1923. They were an in-house SP design, based upon the Baldwin class P-8 and P-10 4-6-2s, but in this case, Alco was the low bidder. Eighteen more class Mt-1 engines followed in spring 1924. The 4-8-2s were intensively used, but carefully maintained; most remained available into the mid-fifties. They were all modernized over the years, but #4300 retained two important early features all the way until retirement—old style crossheads and no disk main driver. However, both a skyline casing and a corrugated sheet-steel pilot were added to the mix. It is pictured at the Brooklyn roundhouse in May 1947, some years before assignment to Commute duties. *(Don H. Roberts)*

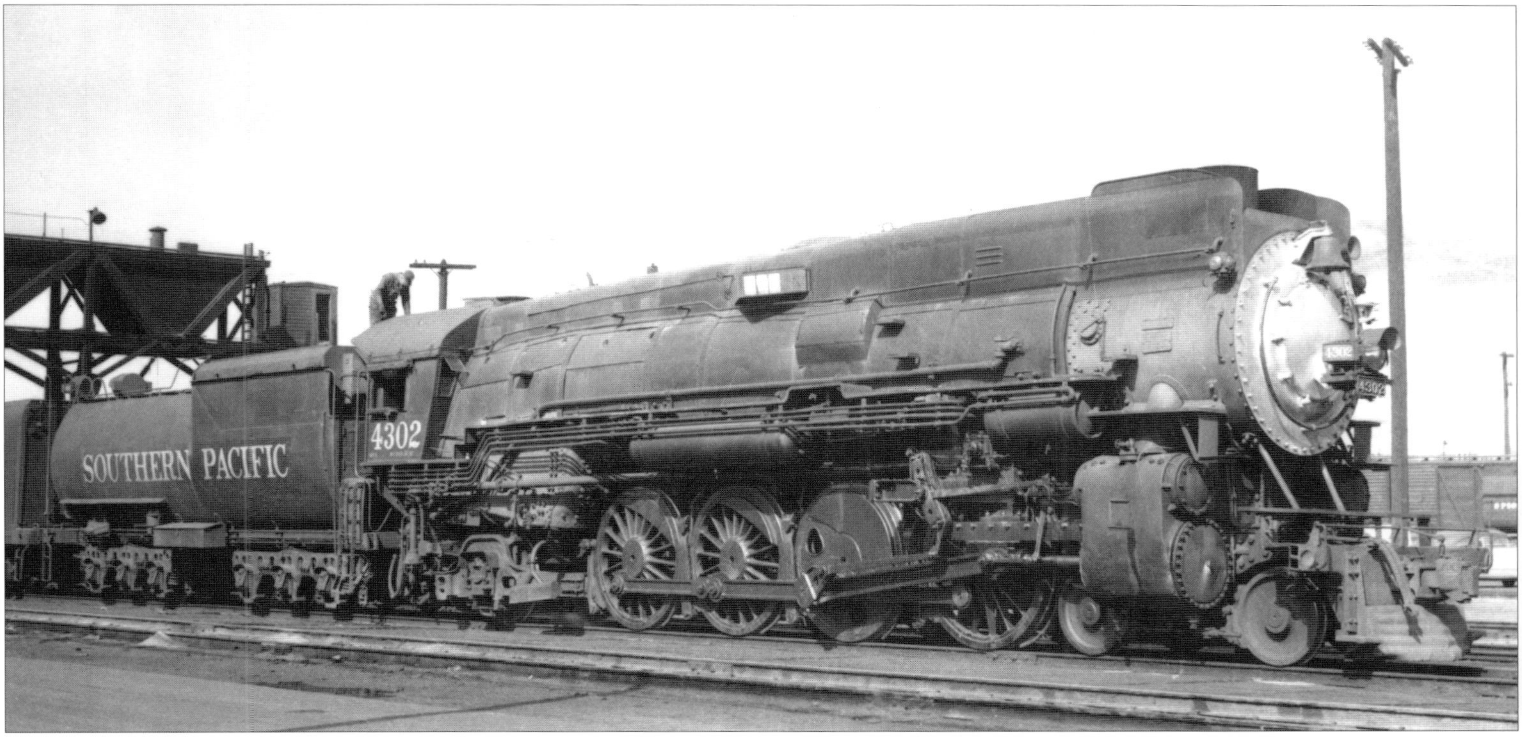

Rio Grande Division assignments for steam locomotives were rapidly dwindling by 1953; within a few months after this photo at El Paso on March 2, 1953, remaining steam power maintained at El Paso will either be transferred to California or retired. Modernized Mt-1 #4302 is pictured on the ready track, featuring a disk main driver, multiple-bearing crossheads, a horn and a corrugated steel pilot. It emerged from Alco's Schenectady works in October 1923 and would be formally "vacated" at Los Angeles about eleven months after this photo. SP would sell it for scrap on June 4, 1954 to the Purdy Company at Los Angeles. *(Grady Robarts)*

All the Mt-1s were built with standard cabs, but were modified in the 1920s with "sport cabs." The second order of Mt-1 locomotives, #4310-4327, entered service in May 1924. Before they were fitted with skyline casings, they looked like this, demonstrated by #4322 at Tracy on December 3, 1939. The #4322 features a 12,000-gallon tender, snowplow pilot and a clamshell smoke deflector (this latter feature installed in December 1937 at Sacramento); it will receive a skyline casing on June 21, 1941 at Sacramento. *(Herman Friedrich)*

By 1954-55, SP's scrapping program for large steam locomotives was in full stride—no longer would a locomotive be carefully stripped of useful accessories before being cut up. By now, physical dismantling was performed by on-line commercial scrap yards near San Francisco and Los Angeles. Only a few splashes of "blackout" paint and a torched piston rod advertise that the #4316 will never be fired up again; even the expensive Universal disk main driver is just so much scrap metal now. The Mt-1 has been dropped off at Dolores yard in May 1955, on its way to scrap at the Purdy Company in Wilmington.

Working all the way from Indio to Beaumont as the point helper on westbound train #39, the *Imperial,* Mt-3 #4336 has been cut off at the summit of Beaumont Hill. Once refilled with fuel oil and water, the 4-8-2 will be turned on the wye and (at the dispatcher's pleasure) make its way back to Indio to await its next assignment. *(Ralph Lettelier: Alan Miller collection)*

Simmering in the afternoon sun at San Jose on October 18, 1953, Mt-3 #4342 will soon be called for a westbound Commute train. Other than an absent trailing-truck booster (removed at Bayshore in June 1953), the #4342 boasts all the visible latter-day modifications: new crossheads, disk main drivers, an air horn and a corrugated steel pilot. It would become one of the last 4-8-2s to be finally cut up for scrap—formally vacated long after steam operations had ceased and sold for scrap on July 13, 1959 to Luria Brothers at South San Francisco. *(Ray Whitaker)*

Pictured in 1946 at the helper terminal in Oakridge, Oregon, Mt-3 #4329 still carried its original nickel-plated cylinder/valve covers as one of the last surviving pampered pets of the Portland Division passenger pool. Shop personnel painted many 4-8-2 smokeboxe doors with aluminum long before tender lettering was revised to the large "Southern Pacific." Latter-day modifications include a disk main driver, 16,000-gallon tender and a corrugated steel pilot.

Eastbound from Los Angeles at El Monte Tower (a Pacific Electric crossing) on June 1, 1941, heavy freight train second #824 required two 4-8-2s. The road engine is Alco #4311, a 4-8-2 that has not yet received a skyline casing and still leads an older 12,000 gallon tender. Point helper #4339 was fitted with a skyline casing the previous October and still carries its bell under the smokebox—in contrast to postwar practice. Even in 1941, selected 4-8-2s had aluminum-painted smokebox doors. *(Phillips C. Kauke)*

Thirteen months after the opening image for this chapter, SP's *Owl* was late again on July 2, 1951, intruding upon the track time of Santa Fe F7 #235 operating on that road's *SCX* train. The diesel-powered freight waits at Cliff in bright sunlight as #4352 (which had lost its *Daylight* paint) finally passes with train #58. *(Phillips C. Kauke)*

It's 2:41 p.m. at the Portland Union Station in the 1946-49 era, before advent of the *Shasta Daylight*; Mt-4 #4348 is leading 4-8-4 #4460 in anticipation of taking the late-afternoon *Beaver* south to Eugene. That jaunty installation of a slanted external whistle (with an airfoil) was not commonly seen. Both multiple bearing crossheads and a pair of disk main drivers have been installed. On the rear of the tender, below the word PACIFIC, there is a shelf-like shield installed to keep overflowing tender water out of the journal boxes.
(Don H. Roberts)

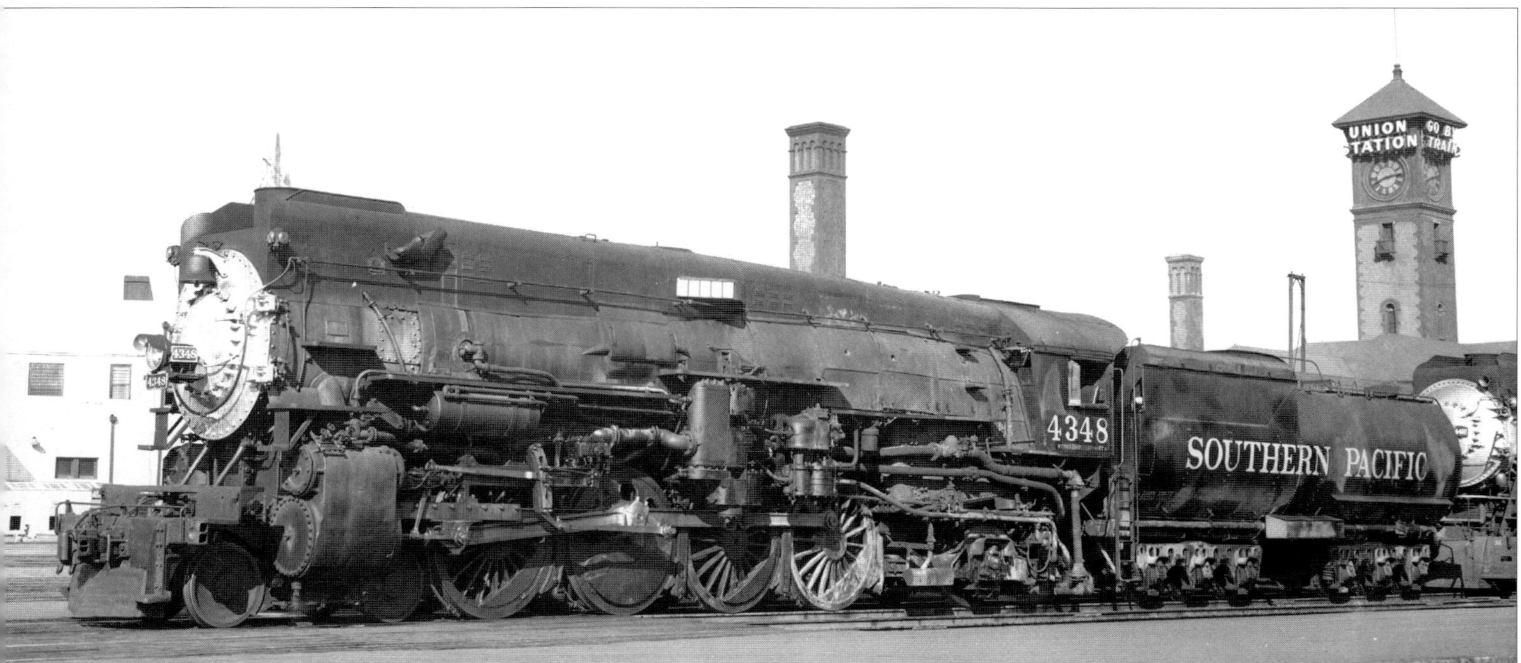

A classic image of the westbound *San Joaquin Daylight* illustrates this heavy train departing Los Angeles in March 1946 with Mt-4 #4352 and GS-4 #4437. Still in the future for this Mt-4 are large tender lettering and then *Daylight* paint on the tender tank and cab. For the moment, however, #4352 features its original pilot and nickel-plated cylinder/valve covers—and an early application of an air horn. *(H. L. Kelso)*

Some eighteen months after the picture above, the #4352 now sports a disk main driver, new crossheads, a corrugated steel pilot and red-and-orange paint. It's the late morning of September 1, 1947 at Lancaster in the Antelope Valley, and train 51, the *San Joaquin Daylight,* has worked through Soledad Canyon from Los Angeles and its locomotives require more boiler water before dealing with Tehachapi. The train will move forward to allow the road engine (also a 4-8-2) to be watered as well.

Still more than two years away from receiving its skyline casing (on August 26, 1949 at Los Angeles General Shops), Salt Lake Division Mt-5 #4370 waits on call for the next passenger train at Sparks on June 2, 1947. The #4370's bell is mounted on a pedestal in front of the smokebox, rather than hanging from a frame as seen on other pictures. *(D. S. Richter)*

specific locomotives to important trains, Mt-4s #4350, 4352, 4353, 4361 and 4363 and their tenders were repainted in red and orange for service on the *San Joaquin Daylight*. All five were repainted in black in 1951.

Discontinuance of long-distance passenger trains and arrival of new diesels meant that SP had less work for the 4-8-2s; many could eventually be found powering freights. Others eventually worked their way to the San Francisco-San Jose Commute Pool, where they supplanted 4-6-2s. Their ample boiler horsepower and tender boosters allowed SP to accelerate its Commute schedules. As late as the fall of 1956, 4-8-2s were still powering Commute trains and local freights in the Bay Area.

The Mt-2s were retired between 1951 and 1953. Mt-1 #4317 was scrapped at El Paso in November 1952 as the first SP-design 4-8-2 to leave the roster. The others were vacated between 1953 and 1956, surviving as hulks after the official end of steam operation. As SP's steam servicing facilities were retired, the 4-8-2s were hauled off to scrap.

SP maintained most of its Los Angeles passenger steam locomotives at the Alhambra Avenue roundhouse near Union Station, not at Taylor Yard—which suggests that Mt-5 #4375 (photographed on the north approach to Taylor roundhouse on the afternoon of June 13, 1947) was probably in freight service on this occasion. Despite modern tender lettering, the #4375 still sports old-style crossheads and pilot and spoked main drivers; unlike #4348 on the previous page, it has no visible whistle at this point in its career. *(Allan S. Youell)*

Two views of #4376 in Commute service in 1955-56, at its final stage of modification, with a corrugated steel pilot, disk main driver and a class 120-C-8 tender handed down from a 4-10-2.

Above: Inbound at South San Francisco on the weekend-only train 165 at South San Francisco early in the morning on a summer Saturday in 1955. *(Charles Givens)*

Below: One of the earlier afternoon trains, weekdays-only 120 departed San Francisco at 4:05 p.m., ahead of the real rush that began in another hour. Seven coaches painted green and one already repainted to two-tone gray trail #4376 out of Third and Townsend on the afternoon of June 18, 1956. It is the final summer for steam-powered Commute trains. *(James J. Buckley: Alan Miller collection)*

On November 1, 1946, not long after El Paso General Shops had rebuilt it with Walschaerts valve gear, multiple-bearing crossheads, corrugated steel pilot and disk main drivers, Mt-2 #4389 rolled in to Colton on the point of train 47, the westbound no-name passenger local between El Paso and Los Angeles. This slow train provided mail and express service with one or more rider coaches for local passengers (and pass-carrying employees) who were not in a hurry. *(Leo Munson collection)*

For decades, the Bay Area's most important athletic contest was the annual Big Game played by the football teams of Stanford and the University of California. Every other year, the teams would meet at Stanford's stadium near Palo Alto, with SP extra trains (from both San Francisco and Oakland) delivering thousands of fans, then taking them home again. For photographers, it was a rare Saturday opportunity to capture a fleet of long, heavy trains at Third and Townsend. On November 21, 1953, five 4-8-2s crowd into a single image: #4342, 4360, 4311 (with air horns) and #4325 and 4324 without horns.

The six former El Paso & Southwestern 4-8-2s that came to the SP in 1924 as #4385-4390 were somewhat larger, heavier and more powerful than SP's own design. Originally equipped with Baker valve gear, their design had no effect on SP practices; in fact, shop forces labored to bring them into line with SP standards. Five were vacated in 1951 and one in 1953, while SP's own were still earning revenue. Originally El Paso & Southwestern #410; the #4385 (pictured at Roseville in May 1948) was built at Alco's Brooks Works in September 1924. It still retains non-standard "staircases" up front, but even they are now more gently aligned than originally. *(Allan S. Youell)*

In the days before urban sprawl, the San Jose-Santa Clara area was a productive agricultural neighborhood, well-served by the SP (with a little competition from the Western Pacific). As a result, the San Jose roundhouse ready tracks hosted not just Commute engines, but the power for an endless stream of local freights. Around 1950, the lineup includes (left to right) Mt-3 #4342 and Common Standard 2-8-0s #2781, 2592 and 2803—with the tender of #2303 silhouetted on the turntable. *(Richard Holmes: Givens collection)*

Always a favorite for photographers, the *Coast Daylight* was scheduled out of Union Station in Los Angeles at 8:15 a.m., immediately curving to the east to face the rising sun. In 1946, a few hundred yards after highball, SP's premier train is approaching Mission Tower near the Los Angeles River. Mr. Kelso sold this photo to Lima Locomotive Works, which in turn gave it to the Southern Pacific, where it graced the covers of many SP publications. *(H. L. Kelso)*

Chapter Fourteen
General Service: 4-8-4

The GS-1, a product of SP's environment, was really a mild redrawing of the blueprints for the Mt-5, featuring the same 73-inch drivers and almost identical wheelbase dimensions. Usually, designers went to a four-wheel trailing truck to dramatically increase firebox size, but the GS-1 firebox boasted just one more square foot of heating surface than on the 4-8-2! However, major gains were made in both superheat heating surface and in boiler pressure—up from 210 to 250—allowing an inch reduction in cylinder size, while still boosting tractive effort.

Baldwin built Pacific Lines #4400-4409 and T&NO #700-703 between May and July of 1930. Within a year of their arrival, the business climate had so deteriorated that SP "stored serviceable" these new GS-1s until the passengers and head-end traffic came back.

Seven GS-1 engines were sent to the Texas Lines: #4406-4409 became T&NO #704-707 in early 1941; #4400, 4403 and 4404 became T&NO #708-710 in November 1943. The #4401, 4402 and 4405 remained on the Pacific Lines and never carried T&NO numbers. In late 1952-early 1953, all ten of the T&NO GS-1s were transferred to the Pacific Lines to run out their last miles.

The streamlined GS-2, in turn, faithfully followed the specifications of the GS-1, substituting cross-counterbalanced disk drivers and multiple-bearing crossheads. Lima built #4410-4415 in December 1936; they entered service in January 1937. Their new trainsets arrived from Pullman in the following weeks and "America's Most Beautiful Train," the *Daylight*, began service on March 21, 1937.

Success of the new *Daylight* was the undoing of the "small 4-8-4" concept at the SP. With the planned introduction of triple-unit diners and an all-Pullman night train, future locomotives would have a greater challenge with "lightweight" train consists. With the GS-3, SP would leave its past behind and move up to 80-inch drivers, 280 pounds of boiler pressure and a 9 per cent larger firebox—yielding about 5,000 horsepower at 60 miles per hour. The #4416-4429 emerged from Lima in late 1937.

SP's ultimate passenger locomotives, classes GS-4 and GS-5, were built at Lima in two lots: #4430-4449 in April-May 1941 and 4450-4459 between March and June 1942. Boiler pressure was again increased, this time to 300 pounds, while cylinder bore went down half an inch and stroke increased by two inches, increasing net cylinder volume. Starting tractive effort remained about the same, but a GS-4 could produce considerably more power at speed. While a single GS-4 still could not conquer Cuesta without a helper, it could keep the heavy *Lark* right on the advertised through the rest of its run.

GS-5s #4458 and 4459 differed from their siblings only in the application of roller bearings to all axles on the engine

(Continued on page 193)

When the 1930 GS-1 order was placed with Baldwin, four were allocated to the Texas & New Orleans as #700-703, while the other ten were delivered to the Pacific Lines as #4400-4409. In 1941 and 1943, SP sent seven GS-1s to the T&NO, where they became #704-709. As the T&NO dieselized, eight T&NO GS-1s were sent to the Pacific Lines in 1952-53, including the four "native" T&NO engines; two GS-1s were scrapped in Texas. The former T&NO #700 moved to the Pacific Lines on December 4, 1952 and was renumbered SP 2nd #4403 on March 16, 1953 (however, this El Paso photo is dated February 26, 1953). On June 24, 1953, it will be renumbered to 2nd #4470. Its next number is shown on page 192. *(Grady Robarts)*

SP was so pleased with the Lima-built class 220-R-1 rectangular tenders for the streamlined GS-2 locomotives that it ordered ten duplicates from Baldwin for use with the GS-1 engines—increasing fuel oil capacity by about 1300 gallons and water storage by about 5,000. Baldwin built 4-8-4 #4408 for the Pacific Lines in July 1930; it was leased to T&NO as their #706 from May 6, 1941 to January 20, 1953. In fresh paint and again wearing its original number, the #4408 was photographed at Los Angeles on October 18, 1953; it would serve until June 1955. *(G. M. Best)*

In the short span of a few months in 1953, SP had two locomotives numbered 4470 on its roster, perhaps simultaneously. The first, originally #4400, worked for nine years on the T&NO as their #708. In March 1953, it returned to the Pacific Lines as *first* #4470, but that July was given back its original SP number. To read the SP records, second #4470 acquired that number in June 1953—before the first was renumbered. In any event, the engine pictured at Taylor Roundhouse on September 30, 1953 is second #4470, briefly second #4403 and originally T&NO #700. *(G. M. Best)*

GS-1s #4401-4402 and 4405 were never transferred to the T&NO, retaining their original numbers. Perhaps the result of upgrades applied after a 1932 boiler explosion, the #4402 was kept in passenger service on the Coast Division to the end; it became the last GS-1 to be used in the Commute Pool. Only a few months remained for steam-powered Commute trains when the #4402 powered train 131 out of San Jose on the morning of August 10, 1956. *(John E. Shaw, Jr.)*

and the tender. The #4458 rolled on Timken, while the #4459 featured SKF bearings. They were the only new SP locomotives built with roller bearings.

In 1943, SP attempted to order sixteen 4-8-4s through the War Production Board, but was rebuffed on its first attempt. The Board finally relented, but only if SP would accept a reduced quantity and revert to the GS-2 design. With traffic to move and no other options, SP was forced to share six 4-8-4s with the Western Pacific. The GS-6 was an un-streamlined version of the GS-2, with two significant changes: a ten-pound boost in boiler pressure and a vestibuled cab. Lima erected #4460-4469 in July and August 1943.

Nine years later, in August 1952, SP leased ten freight 4-8-4s from the Cotton Belt to relieve a power shortage on the Rio Grande Division. They were built at Baldwin in 1930 and at the railroad's Pine Bluff Shops in 1942. The Cotton Belt was virtually all-diesel by 1952, but these low-drivered machines had miles left in them. A test run of the #818 in May 1953 convinced the SP that they could be used in Commute service; in June 1953, they were given SP numbers as classes GS-7 (#4475-4481) and GS-8 (#4485-4487). They were equipped with roller bearings on all axles, but even with disk drivers they were not adequately counterbalanced for passenger train operation. Early trials on Commute trains brought complaints from SP's vociferous customer base and the Cotton Belt engines retreated to freight service. Oddly enough, one more 4-8-4, SSW #813, was renumbered to SP #4488 on February 29, 1956, becoming the last large steam locomotive added to the roster.

Colorful *Daylight* paint was an early casualty of World War II; earlier streamlined GS engines were painted black, then had their sheetmetal skirts removed. GS-4s #4430-4447 lost their streamlining and bright colors in 1950, while #4448-4459 carried on in full *Daylight* paint until 1955. In that year, SP dieselized the *Coast Daylight* with new E9 locomotives and promptly de-skirted the remaining 4-8-4s.

GS-6 #4462 was wrecked in 1953 and became the first Lima 4-8-4 to be scrapped (in March 1954), but as early as 1953 some GS-1s were being laid aside. SP ran off their remaining miles in freight service and on the Commutes, but the economies of full dieselization were too great to resist. The last revenue operation used the #4430 on January 22, 1957. Subsequently all were scrapped except for #4449 (donated to Oaks Park in Portland) and #4460, donated to the National Transportation Museum in St. Louis.

The modern Los Angeles Union Passenger Terminal opened in May 1939, but GS-2s were working out of Los Angeles as early as January 1937—so for more than two years, the streamlined *Daylight* originated and terminated at the old SP Central Station near 5th and Alameda Streets. This picture was taken in the Central Station yard after GS-2 #4411 had delivered its train and train number indicators taken down.

The GS-2 locomotives were painted black (retaining their skirting) during World War II and were never returned to Daylight colors after the war—there are no known images of red-and-orange GS-2s with large "Southern Pacific" tender lettering. The black-painted skirts were removed in 1950 to simplify maintenance. At El Paso on November 7, 1947, skirted #4414 illustrates black paint and new lettering. *(Leo Munson collection)*

The 4-8-2s and 4-8-4s in classes GS-1, GS-2 and GS-6 had 73-inch drivers and a twenty-foot rigid wheelbase; they could comfortably negotiate the tight curve on the east rampart of the Steel Bridge into Portland Union Station. While the later GS engines with 80-inch drivers and a twenty-one-foot-six-inch rigid wheelbase could also get around this tight corner, they were not routinely assigned to the Portland Division. Gleaming in fresh paint and new-style lettering, black-skirted GS-2 #4413 was photographed on March 20, 1946 at the Brooklyn roundhouse. *(Don H. Roberts)*

In the 1946-1950 era, there were still sufficient passenger and accommodation trains in the timetable to keep the 4-8-2 and 4-8-4 locomotives busy, but passenger diesels were making inroads and there was less need for fully streamlined 4-8-4s. GS-2 #4413 went through the West Oakland shop in March 1950 and had its side skirts removed at that time; in the 1950s, pilots painted solid aluminum became the *de facto* standard for non-streamlined Lima 4-8-4s. Cut off from train 19, the Portland-Oakland Mail Train once known as the *Klamath*, #4419 stops for water at Gerber on June 10, 1951. *(Fred A. Stindt)*

First 429, westbound from Tracy, works downhill in Niles Canyon crossing Niles Canyon Road and Alameda Creek, just west of the Dresser spur near milepost 31. It's the early 1950s—a time when SP owned an excess of steam passenger locomotives and ran off their remaining miles on local freight trains. *(Charles Givens)*

GS-3 locomotives (Lima 1937) were painted all-black during World War II and lost their streamlined skirts in 1950; they were still considered first-line passenger power, but deliveries of passenger diesels limited their opportunities. Passenger extras and mail trains were more often their lot as SP's roster of Alco PA locomotives expanded. On October 30, 1954, GS-3 #4425 is powering an eastbound passenger extra at Glendale. *(Stan Kistler)*

The relatively conservative GS-2 design (really a makeover of Baldwin's 1930 GS-1) was quickly followed by the more advanced GS-3, which featured an extended combustion chamber, larger drivers (80-inch *vs.* 73-inch) and longer-stroke pistons. Lightweight alloy steel held down the weight on drivers, while boiler pressure was upped to 280 from 250. Lima delivered fourteen, #4416-4429, between October and December of 1937. At age three, on December 29, 1940, GS-3 #4416 powers the eastbound *Sunset Limited* at Colton. *(Allan S. Youell)*

Once Los Angeles Union Passenger Terminal opened in May 1939, the three tenant railroads were quick to arrange advertising photo shoots from Terminal Tower, a natural vantage point. Unfortunately for the photographer, a Santa Fe train obscured part of this gleaming *Daylight* consist as it departed precisely on time. By a quirk of fate, Santa Fe 4-8-4 #3751 is still available for service in 2010, so it is conceivable that this scene could be replicated today, substituting similar preserved SP engines. *(Jim Harrison collection)*

Outbreak of World War II and fears of Japanese attacks on the California coast prompted SP to "black out" its locomotives, minimizing un-needed light "spill" with sheetmetal light dispersion shields over the headlight and classification lights—presumably, the oscillating headlight was simply disconnected. GS-4 #4439, placed in service on April 15, 1941 (pictured at Glendale on the westbound *Coast Mail*), received a headlight dispersion shield in early 1942. *(G. M. Best: Nestell collection)*

The *San Joaquin Daylight was* Southern Pacific's last main line name train to be powered by steam (at least between Oakland and Bakersfield, if not all the way to Los Angeles). In fact, during the final months of the steam era, passengers on this train were also treated to the use of a full-length dome car. Westbound at Tracy on April 21, 1956, train 51 is in the care of booster-equipped GS-4 #4431. *(Ray Whitaker)*

GS-4 #4436 poses for the Lima Locomotive Works photographer in April 1941; this engine would enter SP service on April 28, 1941. Striped wheel tires were not SP standard but were routinely applied by the builder, especially for publicity photographs such as this.
(Lima Locomotive Works)

Eight years and six months after emerging from the builder's plant, GS-4 #4439 has been called for the eastbound *Imperial,* train 40, and is pictured at Colton on November 6, 1949. Two significant visual changes have occurred during the #4439's short life: the train number indicators have been moved back on the skyline casing and large SOUTHERN PACIFIC lettering has been applied to the tender. In April 1950, the side skirts will be removed, but #4439 will operate for a while afterward with red-and-orange paint still in place.
(Allan S. Youell)

Even after SP's inter-city passenger trains were dieselized, San Francisco-San Jose Commute train service was predominantly steam-powered. The winter of 1956-57, however, would mark the last call for steam passenger power; as a result, during the afternoon rush, considerable film was expended on the sunny side of the Third and Townsend Streets station in San Francisco. Late on the afternoon of July 24, 1956, GS-4 #4437 has backed on to train 144 and waits its turn to sprint down the Peninsula.
(D. S. Richter)

While not as busy overall as Bakersfield, on a daily basis the Santa Barbara station hosted more photogenic trains: both the *Morning Daylight* and the *Noon Daylight,* as well as the *Lark,* mail trains and the *Overnights.* An early casualty of SP's anti-passenger campaign, train 97, the *Noon Daylight,* changes crews and adds water at Santa Barbara in January 1948. *(F. J. Peterson)*

Below: There was a time when every SP main line had its own mail train. Perhaps the best documented was the *Coast Mail,* trains 71-72, that operated between San Francisco and Los Angeles, stopping at virtually every open agency along the way. Steam powered this train until at least 1955, when SP finally gained an oversupply of diesels fitted with steam generators. That occasion was still in the future on March 14, 1954, when eastbound 72—with GS-4 #4440 in charge—stopped at Guadalupe. *(John C. Illman)*

In 1954 and 1955, when SP was just testing the notion of piggyback service (using its own trailers exclusively), well-maintained steam locomotives were available in ample supply—providing an excuse for 4-8-4 power on the *Advance Zipper,* pictured pulling out of the Shops area at Taylor Junction on April 14, 1955. The #4452 lost its skirts and festive paint only a month or so before this occasion.
(John E. Shaw, Jr.)

SP documents report that the #4444 was shorn of streamlined skirts at El Paso on January 23, 1950. With its tender and cab still painted in *Daylight* colors on April 16, 1950, the big 4-8-4 performs a run-by near Colton for photographers on a railfan excursion. At least the #4439 and 4444 were operated with red-and-orange paint after their skirts were removed, but no information has survived on how long the colors actually survived before being painted over. *(F. J. Peterson)*

Nothing but Common Standard-profile heavyweight cars stretching off into the distance distinguishes the first section of train 20, the Oakland-Portland mail train once advertised as the *Klamath*. The date is June 1949 and GS-6 #4461 has crested the summit at Crescent Lake, Oregon. From Odell Lake at milepost 532 this train will have a relatively easy run, with a gentle track profile ahead all the way to Klamath Falls. *(Bert H. Ward)*

Left: This is the only picture the author has seen illustrating a "full colors" GS 4-8-4 in dead storage. There are many later images, taken after the last dozen (#4448-4459) lost their streamlined skirting, but this photo appears to be unique. The #4453 was stored at Bakersfield on April 2, 1955, after dieselization of the *Coast Daylight* in January. SP records indicate that it would be de-skirted on May 16, 1955 at the San Jose roundhouse. *(R. D. Patton)*

Below: The ten GS-6 locomotives, #4460-4469, were built to GS-2 blueprints with the addition of a vestibule cab. By 1943, during World War II, lightweight steel was in short supply; as a result, a GS-6 weighed ten tons more than its prototype. Never fully streamlined, the GS-6s were widely used on secondary trains and as first-line power on the Shasta and Portland Divisions. The #4466, built in August 1943, powers mail train 25 at Berkeley on June 28, 1947. *(Guy L. Dunscomb)*

One of only two SP steam locomotives purchased new with roller bearings, GS-5 #4458 was fitted with Timken equipment, while the #4459 sampled the SKF product. Both went to work in June 1942 and proved successful—at least to the extent that SP never chose to rebuild them with standard bearings. From their arrival, the last two new SP steam passenger engines were operated routinely as part of the Coast Line passenger pool. On September 10, 1951, the #4458 has brought the eastbound *Coast Daylight* from Santa Barbara, making a brief stop at suburban Glendale (five miles out) before terminating at Los Angeles. *(Allan S. Youell)*

Two hundred fifty mileposts from San Francisco, train 52, the *San Joaquin Daylight*, arrives at Tulare on May 14, 1955. Roller bearing-equipped GS-5 #4459 lost its red-and-orange paint a few months before this picture, but steam will power the Valley's premier passenger train for another year. Alco RSD-5 #5339, less than two years old, is parked in the clear. The Alco diesel will serve only for seven years before being traded back to its builder, while the Lima 4-8-4 would work for fourteen. *(John E. Shaw, Jr.)*

With ample weight on the drivers and trailing-truck boosters, the ten never-streamlined GS-6 locomotives were kept busy powering heavy trains on mountain districts. One of the two regulars on the Portland Division (the other was #4469), the #4462 is pictured on May 29, 1948 at the Brooklyn roundhouse across the Willamette River from downtown Portland. This engine would be wrecked beyond economical repair in 1953 and became the first Lima GS to be scrapped. *(Don H. Roberts)*

While the SP occupied Altamont Pass first, there was plenty of land among these gentle hills for a second railroad—the Western Pacific—to lay competing rails. In fact, the location of this particular photo, illustrating extra 4465 west on July 16, 1955, can best be defined in WP terms as "near Redmond Cut." The GS-6 is working off its last miles as power for Western Division freight locals. *(Stan Kistler)*

"Lighweight" car construction, at least as interpreted by the Southern Pacific, seemed to be a blank check to add heavy features to its streamlined trains. Triple-unit diners and heavyweight head-end cars meant that the *San Joaquin Daylight* would require ample power to deal with both Soledad Canyon and Tehachapi, even though a single locomotive could do the job down in the Valley. About 1950, GS-6s #4460 and 4463 doublehead on train #51 in Soledad Canyon. *(Ralph Lettelier: Miller collection)*

Non-streamlined and lacking many of the innovations applied to the GS-4 and GS-5 locomotives, the GS-6 machines (#4460-4469; Lima, August 1943) were literally a throwback to SP's 1936-vintage GS-2s. Military needs during World War II usurped the supply of lightweight steel, so many moving parts and the boiler weighed more than if they had been manufactured before the war. However, one great boon to enginemen on cold nights in the Cascades was the addition of a GS-4 style vestibule cab. The #4464 is pictured on August 27, 1948 at the Brooklyn roundhouse and will serve until the end of SP steam in 1956.
(Don H. Roberts)

Returned to its roots as a freight engine, SP #4481 is pictured in 1955 on eastbound San Joaquin Division freight train 780 between Slater and Famoso. One of seven Baldwin-built class GS-7 locomotives leased from the Cotton Belt in 1952, the #4481 began its career in October 1930 as St. Louis Southwestern #809. Time was running out for this class; the #4481 would be vacated in October 1955. *(J. R. Knoll)*

SP soon ran out of assignments for the former Cotton Belt GS-7 and GS-8 locomotives; their day in the California sun would be cut short by a flood of new GP9 units from EMD. Winter traffic doldrums in 1955 parked most of the remaining serviceable SP steam locomotives, but they were set aside with great care in anticipation of running again during the 1956 rush. Clean, lubricated and available, #4476 (SSW #804; Baldwin 1930) is stored at Los Angeles on December 3, 1955. *(John E. Shaw, Jr.)*

In the summer of 1952, the Rio Grande Division was running short of locomotives; however, the Cotton Belt had serviceable 4-8-4s in storage, so ten were leased to the SP in August 1952. They were lighter than the GS-2s and were designed for freight service (with 70-inch drivers), but they were willing workers on SP freights in the New Mexico desert. Initially, they retained their SSW numbers and lettering, but were treated to train number indicators and aluminum-painted smokebox doors. Cotton Belt #807 (Baldwin 1930) is pictured on March 13, 1953 at El Paso; it will become SP #4478 on June 15, 1953. *(Grady Robarts)*

One occasion when a foreign-line locomotive was used on a Commute train found Cotton Belt 4-8-4 #818 on train #126. The date was May 27, 1953; this successful test resulted in SP moving nine additional leased SSW 4-8-4s to California in June and July 1953 and renumbering them 4475-4481 and 4485-87. Blank numbers were left in case SP decided to draft more 4-8-4s, but only SSW #813 came west later, becoming SP #4488 in February 1956. While powerful and capable of good acceleration, these machines were designed for freight service and gave SP commuters a rough ride. *(B. A. Romano)*

Train 55-56, the *Tehachapi,* operated between the Oakland Mole and Los Angeles from 1936 to 1955. Once the *San Joaquin Daylight* began service on this route, the *Tehachapi* lost its status as a named train and evolved into the San Joaquin Valley's mail train. On this 1954 occasion at Madera, the all-heavyweight westbound consist has dwindled to five head-end cars and a rider coach. Locomotive #4475 is the former Cotton Belt #802, renumbered June 9, 1953, the lowest-numbered GS-7. *(Guy L. Dunscomb)*

Even though they were equipped with roller bearings, the former Cotton Belt 4-8-4s were not equalized for passenger train speeds and delivered a "pulsating" thrust that irritated commuters. At first, SP added an empty buffer coach to Commute trains powered by GS-7 and GS-8 engines, but the ultimate solution was to return these low-drivered engines to freight service. In the summer of 1955, the first section of westbound freight 423 pulls away from the Berenda tank after #4485 has taken water. Berenda is 7.4 miles North (railroad East) of Madera on the San Joaquin Valley main line. *(Charles Givens)*

The last large steam locomotive added to the SP roster did not debut until 1956. SP #4488 was renumbered from Cotton Belt #813 on February 29, 1956, serving briefly in the Bay Area and putting on a few more miles in the San Joaquin Valley. Almost certainly, it ran up less revenue miles than any other modern SP locomotive. Built by the SSW's Pine Bluff Shops in Arkansas, this machine featured roller bearings on all axles; it is pictured on March 10, 1956 at Los Angeles, only days after assuming its final identity. *(John E. Shaw, Jr.)*

A longtime SP tradition was the notion of operating local and "junk" freight trains over a single division. Once a train reached a division point, it might assume a new identity and continue on—or be broken up and its cars forwarded in many different directions. Westbound (geographically more north than west) freight 421 is powered by GS-7 #4480 departing the Fresno Yard in the dead of a San Joaquin Valley winter on February 20, 1955. *(John E. Shaw, Jr.)*

Cotton Belt #817, home-built in November 1942 while SSW waited for its FT diesels, was a low-driver freight engine erected at the railroad's Pine Bluff Shops. In 1952, SP leased it for freight service in New Mexico, and then transferred it to California for Commute duties, where it was renumbered to #4486 on June 17, 1953. Passenger service didn't last long; the GS-7s and GS-8s were freight haulers, not smoothly-accelerating passenger engines. In October 1953, the #4486 departs with train 146 from Third & Townsend, San Francisco. *(W. C. Whittaker)*

With small, 63-inch drivers and significant counterbalancing problems that led to a rough ride at speed, the 4-10-2s were not considered first-line passenger engines—but they did offer sufficient power so that SP could avoid doubleheading some of its slower accommodation trains. For example, the all-heavyweight (and never streamlined) *Argonaut,* train 5, was typically assigned a 5000 at Indio for the climb over Beaumont Hill into Los Angeles. This heavy train was not scheduled to push #5011's fifty-mile-per-hour speed limit, even on relatively level track west of Colton. Crossing the Los Angeles River and passing Mission Tower on the afternoon of November 18, 1945, the *Argonaut* will reach the Union Passenger Terminal station platforms a few seconds after this photo. *(H. L. Kelso)*

Chapter Fifteen
Southern Pacific Type: 4-10-2

In the 1920s, the SP began exploring new developments in steam freight locomotives in hopes of making significant technological leaps from its legacy of the Harriman era. The 2-10-2 was developed and refined between 1917 and 1925, but reached its practical limits with the F-5 class. Alco, however, armed with the Gresley valve gear patent rights, was by then touting its three-cylinder concept and proposed a ten-driver locomotive with exactly the same 22-foot-ten-inch rigid wheelbase (including a lateral driver box on the front driving axle) as the Baldwin 2-10-2s, but with three cylinders and about 10% greater boiler capacity. Alco offered a 12% power increase with just 3% additional weight on the drivers.

SP signed for sixteen, #5000-5015, for delivery in 1925 and followed up with orders for thirty-three more (#5016-5048) in 1926 and 1927. They were given the name "Southern Pacific Type" and posed for numerous publicity photos. The Union Pacific (Los Angeles & Salt Lake) took delivery of nine slightly smaller 4-10-2s (#8800-8808) in May and June 1926, contemporary with the Southern Pacific's class SP-2, but the LA&SL engines were erected in Dunkirk, New York, at Alco's Brooks Works, while the SP's emerged from Schenectady.

Unfortunately, SP purchased a significant fleet without evaluating a prototype in real world conditions. They spent too much time in the shop (as did the similar Union Pacific machines) and damaged rails. A ten-driver locomotive was not well suited for The Hill; within a short time they were reassigned to the Los Angeles, Tucson and Rio Grande Divisions, where they were to remain until after World War II. Boxpok disk main drivers and cross-counterbalancing were installed during the 1940s, allowing a 10% increase in maximum speed to 55mph. By 1940, all of the 4-10-2s were assigned 16,000 gallon water-bottom Vanderbilt tenders.

The Union Pacific's small group of three-cylinder 4-10-2 locomotives were rebuilt during World War II to two cylinders, but the SP stood pat; about the only major modification program was a retreat from trailing truck boosters that saw thirty 5000s lose these devices in the 1946-52 era at the shops at El Paso, Tucson and Los Angeles. Many were fitted with clamshell smoke deflectors, a second sand dome and corrugated steel pilots.

With the 1940s demise of the 3900 and 4000 series cab-ahead locomotives, SP decided to transfer about half of its 4-10-2 fleet to Oregon to take over Willamette Valley shuttle freight service between Brooklyn Yard and Eugene. Included were the #5000, 5002-03, 5005-08, 5011-13, 5017-18, 5024-25, 5027, 5031-34, 5036-37, 5044 and 5046-47. Beginning in March 1953 and ending in December 1954, the Portland Division disposed of 24 4-10-2 locomotives—its entire fleet. The remaining 4-10-2s were scrapped in the same era at Sacramento or El Paso.

Class SP-2 #5021 became the last large steam locomotive on the Portland Division, and was finally vacated in October 1955. Presented to the Southern California Chapter of the Railway & Locomotive Historical Society, it remains on display at the Los Angeles County Fairgrounds in Pomona.

This SP-1 entered service in July 1925, serving through World War II and after on the Los Angeles Division—where water was relatively precious and baths for freight engines few and far between. Southern District locomotives were frequently photographed with SP's characteristic "clamshell" exhaust deflector; the #5002 (pictured at Los Angeles in November 1948) also carries the modern corrugated solid steel pilot and a disk main driver. It will soon be displaced by new diesels and move to Oregon to run out its last miles.
(Herman Friedrich)

Afternoon sunlight on the outbound ready tracks at Taylor Roundhouse (a few miles north of downtown Los Angeles) provided a favorable setting for countless SP locomotive photographs. SP-1 #5003 has been serviced and is ready for its next assignment—probably up the coast—on October 31, 1948. By the early 1950s, it will be working in Oregon. *(Allan S. Youell)*

Southern Pacific's first 4-10-2, the #5000, entered service on June 20, 1925. It was apparently the last class SP-1 locomotive in the Bay Area by the time this picture was taken at Santa Clara in March 1953 and would soon be transferred to the Portland Division. The big three-cylinder 4-10-2 escorts a perishable block headed east by way of Oakland and the Overland route. SP apparently had planned to save the #5000 for some future display, as it would be vacated at Brooklyn (Portland) in May 1954, but held until November 1956 (the last 5000 to leave the property), when finally sold for scrap to Oregon Steel Mills. *(Charles Givens)*

Many 4-10-2s were assigned to the Portland Division in the years after World War II, replacing old Baldwin 4-6-6-2 and 2-8-8-2 locomotives that had been worn out during the war. The #5010 worked in Oregon during the late forties and early 1950s, but spent its last days in northern California. It is pictured in October 1950, piloting arriving westbound freight train 663 at Eugene from Brooklyn Yard. *(J. T. Boyd)*

Recently painted and benefiting from regular washes, gleaming SP-2 #5024 stops on the lead of the Brooklyn Roundhouse on May 31, 1951. Arrival in 1954 of the first Electro-Motive SD9 locomotives on the Portland Division portended the end of big steam on the SP in Oregon; between March 1953 and December 1954, two dozen 5000s would be vacated at Brooklyn—many of them cut up on the spot. The #5024 was vacated on September 2, 1954 and sold for scrap that November to nearby Oregon Steel Mills.

(Don H. Roberts)

The World War II version of "Homeland Security" has clamped down on SP steam locomotives, especially those operating near the coast. Headlights and even small classification lights have been deeply hooded to prevent unnecessary spillover of light that might provide a target for attacking airplanes. Class SP-3 #5040, which entered service on September 1, 1927, models its blackout features at Taylor Roundhouse in 1942. *(G. M. Best)*

Below: Sixty years ago, Beaumont was a busy turn-back point for many helpers and a welcome break for water service. Centralized Traffic Control was installed between Colton and Indio in January 1944, resulting in a significant increase in capacity on this helper district. This undated photo finds SP-3 #5045 in charge of a reefer block; no doubt there will be helpers spaced back in the train. *(Ralph Lettelier: Miller collection)*

The #5029, an SP-2 placed in service on July 23, 1926, was one of many of these big machines that worked in helper service on Beaumont Hill. It is pictured in low sunlight at Colton on December 7, 1947. Unlike many of its compatriots, the #5029 remained a Sunset Line stalwart; it would be vacated at Tucson (the third 5000 to leave the roster) on January 22, 1953. *(Allan S. Youell)*

As late as 1952 and 1953, before SP owned a sufficiently large diesel fleet to assign diesel helpers to the Sunset Route, Colton and Indio enginehouses stabled ample reserves of 4-10-2s for use in both directions on the gentle curves of Beaumont Hill. Extra 5045 west, at Garnet on August 27, 1952, featured three 5000s, with #5045 leading, and #5041 and another 4-10-2 pushing ahead of the caboose. *(Don Ball: SP Historical & Technical Society)*

In the post-World War II era, the Southern Pacific narrow gauge operation in the Owens Valley was powered by three relatively modern, forty-two ton Baldwin class 10-26D locomotives, built in 1907, 1909 and 1911. They did not become part of the SP roster until 1929, when the SP purchased the Nevada-California-Oregon Railway and replaced older locomotives with newer N-C-O machines. SP considered many proposals to standard-gauge this line, but the outgoing freight generated on-line could be inexpensively transferred to standard gauge cars. Instead of changing the gauge, SP elected to pay General Electric $75,500 for a small diesel unit that arrived in October 1954. SP retired two of the three 4-6-0s (#8 and 18) and donated them for display, retaining the #9 in standby service until the line was abandoned in 1960. Visiting rail photographers provided the excuse for the Slim Princess to show off, as on this February 1952 occasion. *(Arthur L. Lloyd)*

Chapter Sixteen
Narrow Gauge, Shop Switch and 4-4-0

The Southern Pacific operated a narrow gauge "subsidiary" in eastern California, but never bought a new narrow gauge steam locomotive. Six three-foot-gauge ten wheelers operated in 1940 on the truncated California end of the original Carson & Colorado, none of them native to that operation. The #14 and 17, Baldwin 1886 and 1887, were transferred from the South Pacific Coast when that Bay Area line was converted to standard gauge in 1906-07. The other four—two decades younger than the ex-SPC machines—were brought aboard when the SP took over the Nevada-California-Oregon. They weighed between 32 and 44 tons; only the #18, Baldwin 1911, featured Walschaerts valve gear.

As narrow gauge operation contracted during World War II, three 4-6-0s, #14, 17 and 22, were hauled off to provide stationary boiler service at Salem, Colfax and Bayshore. The former N-C-O Baldwins worked into the fifties in the Owens Valley, but in 1954 SP bought a small General Electric diesel unit—allowing immediate retirement of #8 and 18. The #9 remained available as standby power for GE diesel #1 until the line was abandoned.

Large roundhouses always have a need for moving cold locomotives; SP's solution to this problem was to inexpensively rebuild old switchers (in some cases dating from the 1880s) with saddle tanks for fuel and water, eliminating the tender entirely and allowing the shop switcher to fit on turntables with the locomotive it was moving. Shop switchers (as non-revenue equipment) were renumbered into the three-digit portion of the Maintenance of Way number series. Fifteen 0-6-0T shop switchers operated during the period addressed in this book.

While it is presumed that employees at a given roundhouse rebuilt "their" shop locomotives on site on a time-available basis, there is some evidence that shop switchers may have been rebuilt at a central shop—probably Bayshore—and doled out to different locations. In any event, no two were identical and few were transferred between facilities. Not every roundhouse had its own shop switcher; Bakersfield, San Jose and Tracy all did without.

One shop switcher was a true relic: #571 was built as a 2-6-2T suburban passenger engine before the days of the Oakland-Alameda-Berkeley electrification; all that was done to convert it was to remove the front truck! Built at the Sacramento Shops in January 1882 as Central Pacific 2nd #231, it later became SP #1502 and then #1901. It was the last CP engine on the roster, retired in 1947.

The six surviving 4-4-0 locomotives on the Pacific Lines were renumbered in 1938 to #1500-1505, but four were vacated from the roster in 1939-40, leaving only #1500 and 1502. These two survived through World War II in stationary boiler service, but in 1949 and 1951, SP wrote them off.

Countless photographs were taken of Southern Pacific narrow gauge locomotives and trains once railfan photographers returned from their war duties. SP also supported the movie business, providing a true desert environment for many a cowboy flick—drafting mainline management to ensure that all went well. This picture—at the station point of Alico in 1957—proves the point that SP narrow gauge steam locomotives were already "radio equipped" in the 1950s, at least when movie cameras were out of range. *(Strapac collection)*

Built in November 1909, Nevada-California-Oregon #9 retained its original number after moving to the SP in 1929. It would earn the dubious honor of being the very last SP steam locomotive to operate in regular service, assigned to standby duties after arrival of GE diesel #1 in 1954. Vacated in the spring of 1960, the #9 would be donated for display at the on-line Owens Valley village of Laws.

(Arthur L. Lloyd)

After World War II, the Bakersfield backshop assumed responsibility for maintaining the remaining narrow gauge locomotives; a ramp at Owenyo allowed narrow gauge rolling stock to be rolled onto standard gauge flatcars for the 211-mile ride via the "Jawbone" and Mojave to Bakersfield. In 1952, ten-wheeler #9 has a shop date and is already safely blocked aboard a flatcar behind Mikado #3203. Near-twin #18 (the only one of the last three 4-6-0s with Walschaerts valve gear) will answer local calls during #9's absence.

(Strapac collection)

The "Armstrong" turntable at Laws, the far northern terminus of the postwar narrow gauge, was kept in service after official dieselization in 1954. Diesel #1 didn't require turning (and would have probably over-stressed this venerable wooden structure), but Baldwin ten-wheeler #9 was brought out to work when the #1 needed to take a ride down to Bakersfield for maintenance or new paint. On those occasions when the #9 came out, news traveled fast and photographers gathered. It's August 1959 and the SP narrow gauge has less than a year until abandonment. *(Herb Cearley: Miller collection)*

The old Schenectady Locomotive Works built few narrow gauge locomotives, but one notable exception was a group of six 1899-1900 vintage ten-wheelers built for Colorado's Florence & Cripple Creek. These machines, as built, had their cabs located so far forward on their backheads that they were called "deckless;" the engineer and the fireman were relatively isolated from each other. One, the #20, is preserved at the Colorado Railroad Museum. F&CC #22 was sold in 1915 to the Nevada-California-Oregon and became SP #22 in 1929; it was retired to stationary boiler service at the Bayshore roundhouse, where it was photographed in May 1946. *(D. S. Richter: Menke collection)*

In 1939 and 1940, skilled sheet-metal workers at Los Angeles General Shops assembled two similar (but not identical) 0-6-0T locomotives from retired 1903-vintage S-5 tender engines. Unlike all the other SP roundhouse switchers, these two had a touch of "industrial design," featuring a streamlined shell over their plebian saddle tanks. The #565, assigned to the Shops complex on Alhambra Avenue, entered service on August 28, 1939. Following on February 6, 1940, the #567 was posted to the Taylor Roundhouse. Photographed on November 23, 1947, the #567 is posed on a roundhouse lead. *(Herman Friedrich)*

Locomotive facilities with old shop switchers were expected to use them until they wore out, which meant that some outlying roundhouses were switched by real relics into the modern age. Elderly Ogden shop goat #571 (across the page), which had worked there since 1916, finally breathed its last in 1947. A relatively modern—but still not superheated—class S-8 0-6-0, #1155, was converted to saddle-tank operation to switch the Ogden roundhouse. It was photographed there on July 22, 1950. *(Strapac collection)*

With new diesels arriving almost daily in the late 1940s, SP had a surplus of small steam switching locomotives. Most would be scrapped, but a few had good boilers and could be put to work in non-revenue service. Less than two months after this photo (at Roseville on May 31, 1948), SP MW 619 would be set aside. This venerable machine had been the roundhouse switcher here since August 1921, rebuilt from an 1888-vintage, Rhode Island-built 0-6-0. A 1911 Baldwin S-8 (#1191) would be rebuilt to replace the #619. *(Allan S. Youell)*

Rebuilt in late 1928 from 1893-vintage Schenectady 0-6-0 #1041, SP MW 620 served the El Paso Shops and roundhouse. SP's saddle-tank conversions were all generally similar in specifications (varying mainly in precise fuel and water capacity), suggesting that the Drafting Room prepared erecting drawings for each one. Shop employees fabricated sheet-steel tanks, closed up the cab and replaced the rear drawbar with an automatic coupler and an airbrake hose. The #620 posed for this image at age ten and was to survive until until May 1953, as the Rio Grande Division was dieselized. *(G. M. Best)*

While the SP operated numerous facilities on the west (downtown) side of the Willamette River in the Portland area, cheap land on higher ground was available across the river in suburban Brooklyn, so that's where the main yard and locomotive facilities were established. SP MW 566, another former Central Pacific 0-6-2T, served here until 1939. It was replaced that year by class S-5 (Baldwin 1901) #1079, from a group that was then being retired. The #1079 became SP MW 208 and would remain in service until steam departed the Portland Division; it was photographed there in September 1945.
(Guy L. Dunscomb)

Built by the Schenectady Locomotive Works in September 1888 as Southern Pacific of California #308, a standard 0-6-0 with a wooden cab and a kerosene headlight, this venerable machine was no longer useful out in the yards by late 1917. The shop crew at Tucson fabricated the small oil fuel tank and the larger water tank and installed an automatic coupler on the rear—creating SP #MW 564 as the Tucson roundhouse goat. This class S-1 was photographed there in September 1937 and would continue to work until May 1954.
(Guy L. Dunscomb)

Dating from January 1882, Ogden shop switcher #MW 571 was the oldest tank engine still operating on the SP after World War II. Furthermore, it was the last surviving Central Pacific locomotive in service. Built for Oakland commute service as a 2-6-2T (CP 2nd #231), it became SP #1502 in 1891 and then #1901 in 1901. When the SP electrified its local passenger services in the East Bay, the little 2-6-2Ts were thrown out of work. In 1916, the #1901's leading truck was removed and the locomotive demoted to roundhouse duties; twenty years later, it was still at work at Ogden.
(E. B. Kille collection)

Each one of the SP's roundhouse switchers had some unique feature, because each was rebuilt from a retired locomotive, using spare parts and local ingenuity. By far the most powerful of SP's shop switchers, MW 966 was rebuilt in 1937 from second #1010, a former El Paso & Southwestern six-wheel switcher that came to the SP in 1924. It was a fixture at the Bayshore Roundhouse and backshop, home port for locomotives assigned to the Coast Division. Bayshore was located just south of San Francisco City-County limits, in Brisbane. Douglas Richter captured this image of SP MW 966 at Bayshore on September 2, 1954.

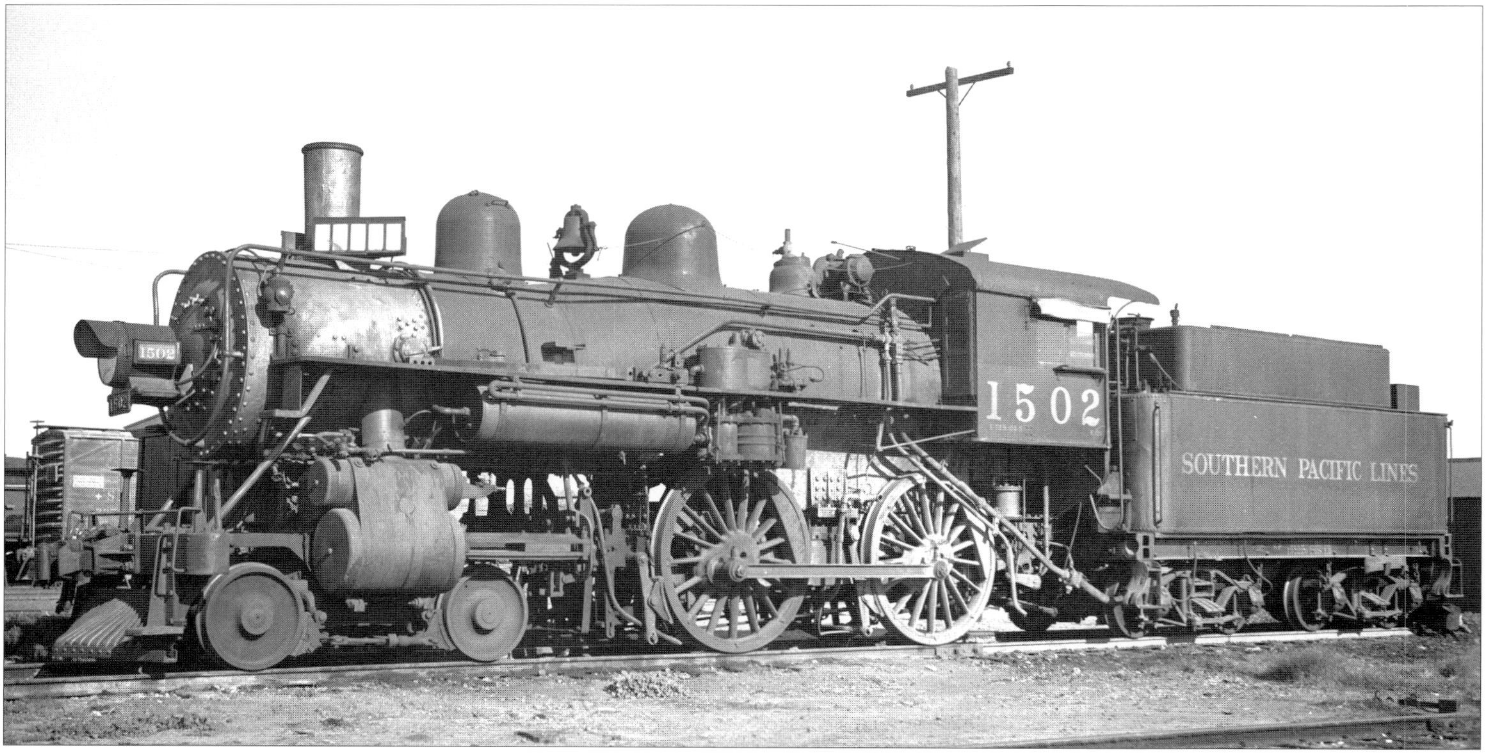

By 1938, SP had very little use for its remaining class E-23 4-4-0 locomotives, built 1898-1900 by Cooke and Schenectady. The last six survivors were renumbered to #1500-1505 in 1938—but four were vacated almost immediately. The #1502 is pictured on November 19, 1940 at Tracy with its main rod removed. By 1945 it would be permanently assigned to stationary boiler duty at Indio and would finally be scrapped in August 1951 as the very last Pacific Lines eight-wheeler. *(Allan S. Youell)*

Two narrow gauge 1880s-era ten-wheelers (#14 and 17) inherited from the South Pacific coast survived into the modern era, retained for use as roundhouse stationary boilers. Their paint jobs reflected current standards, although the #17 (pictured at Salem, Oregon on June 13, 1949) never pulled a train after retirement from active service in December 1945; the tender should not have been lettered SOUTHERN PACIFIC because the locomotive had been vacated prior to the change in lettering standards from "Southern Pacific Lines." The #17 would remain parked on this short piece of narrow gauge track until scrapping in April 1952.
(R. W. Biermann: Menke collection)

While narrow gauge railroads might have been less expensive to build than their standard gauge counterparts, the costs of transferring freight from one gauge to another usually outweighed any savings earned during construction. With most of its freight outbound mineral products from local mines, the SP narrow gauge utilized this labor-saving transfer trestle at Owenyo to dump bulk products directly from narrow gauge gondolas to standard gauge cars below. The date is August 1959 and this is perhaps the very last time that 4-6-0 #9 climbed the ramp to the Owenyo transfer trestle. *(Herb Cearley: Miller collection)*

The last Pacific Lines 4-4-0 locomotive to turn a wheel under its own power was E-23 #1500, built in August 1898 by Schenectady. It received a superheater and piston valves in 1918, doubtless extending its life into modern times. The SP leased it to the Northwestern Pacific from August 1948 to February 1949; even the subsidiary had little use for it and it was cut up at Sacramento that March. It posed for an end-of-an-era portrait on November 13, 1948 at NWP's waterside terminal in Tiburon, across the Golden Gate from San Francisco. *(Fred A. Stindt)*

SP's granite quarry at Rocklin was often switched by the Roseville shop switcher because the light rail there could not support a larger locomotive. While this appears to be a "road job," Rocklin was within the Roseville yard limits, so the switching crew didn't draw extra pay. Usually the shop goat would be towed to Rocklin and back to save fuel and water. In this unique picture, taken in December 1955, much of the Sacramento Valley had been flooded and SP needed all the rip-rap it could get.

(Dave Oliver: Dill collection)

Shop switchers were typically rebuilt from obsolete tender-equipped 0-6-0s, but when conversions were required in the 1940s, SP dipped into its large pool of Common Standard class S-8s (there were never any superheated, piston-valve 0-6-0T engines on the MW roster). Between 1940 and 1948, S-8s #1295, 1162, 1155 and 1191 became SP MW #217, 220, 569 and 572. At the very end of the steam era, in September 1956, West Oakland roundhouse goat MW #217 uses an adapter "coupler" to move a tenderless cab-ahead.

(Harold Beal: Dill collection)